FOOD, MAN, AND SOCIETY

FOOD, MAN, AND SOCIETY

Edited by
Dwain N. Walcher
The Pennsylvania State University
University Park, Pennsylvania

Norman Kretchmer
National Institute of Child Health
and Human Development
Bethesda, Maryland

and
Henry L. Barnett
Albert Einstein College of Medicine
of Yeshiva University
Bronx, New York

Plenum Press · New York and London

Library of Congress Cataloging in Publication Data

Food, man, and society.

"Papers . . . presented at the third meeting of the International Organization for the Study of Human Development, held in Madrid on September 21-24, 1975."

Includes bibliographical references and index.

1. Food—Addresses, essays, lectures. 2. Food supply—Addresses, essays, lectures. 3. Nutrition—Addresses, essays, lectures. I. Walcher, Dwain N., 1915-
II. Kretchmer, Norman, 1923- III. Barnett, Henry Leslie, 1914-
IV. International Organization for the Study of Human Development.

TX353.F663 301.5'2 76-28698

ISBN 0-306-30974-2

Papers presented at the third meeting of the International Organization for the Study of Human Development, held in Madrid on September 21–24, 1975

A Division of Plenum Publishing Corporation
227 West 17th Street, New York, N.Y. 10011

Printed in the United States of America

Preface

The papers contained in this volume were presented at the third meeting of the International Organization for the Study of Human Development, held in Madrid on September 21-24, 1975. The primary objective of the Organization is to bring together persons from a wide range of disciplines concerned with problems of human development. The first meeting typified this interdisciplinary approach by concentrating on problems relating to genetic expression. The second meeting considered milk and lactation from a variety of viewpoints, including those of the molecular biologist and the geographer. The present meeting deals with nutrition as an integral part of human life as well as a fundamental discipline in the area of biology.

The editors wish to thank Professors Ettore Rossi and Fabio Sereni and Dr. Frank Falkner for their invaluable assistance in planning this conference, and would particularly like to express their appreciation toward Professor Cipriano Canosa and his staff for their gracious and hospitable assistance in providing facilities and accommodations for the participants. The editorial assistance of Mary Lynn Hendrix has aided us in our contribution; and the secretarial support of Millie Fazzi and Natalie Stover is gratefully acknowledged.

The continued support of the Nestlé Company, Inc., as evidenced by essential financial contributions and genuine encouragement for the concept of human development has sustained the organization in its developing years. Similar support has been given by The National Institute of Child Health and Human Development. The interest and financial support of Ross Laboratories is genuinely appreciated.

D.N.W.
N.K.
H.L.B.

EDITORS' AFFILIATIONS

Dwain N. Walcher, M. D., Director
Institute for the Study of Human Development
College of Human Development
The Pennsylvania State University
University Park, Pennsylvania

Norman Kretchmer, M. D., Ph.D., Director
The National Institute of Child Health
and Human Development
Bethesda, Maryland

Henry L. Barnett, M. D.
Professor of Pediatrics
Albert Einstein College of Medicine of
Yeshiva University
Bronx, New York

Contents

Introduction

In most societies, throughout most of history, few individuals have been able to escape from an obsessive concern with food long enough to undertake a systematic study of any other subject. All of the contributors to this book, and presumably all its readers, are among the fortunate minority who have escaped. As a result of having been born into societies which have achieved a high degree of agricultural productivity, we have been freed from the drudgery of subsistence agriculture, and from the endless search for food which shapes the life of the hunter-gatherer.

Our liberation began with the domestication of plants and animals about ten thousand years ago. This event, or series of events, led to a dramatic increase in the efficiency of food production, which in turn permitted the establishment of the complex, densely populated entities known as civilizations. In a sense, then, civilization, and the arts and sciences which constitute its highest achievements, are by-products of agricultural progress. The highly differentiated academic disciplines which contribute to our understanding of nature and civilization could never have arisen in the absence of an efficient agricultural system.

Thus it seems altogether fitting that the current world food crisis should bring together representatives of the various disciplines in a collective effort to achieve a deeper understanding of the age-old triangular relationship between food, man, and society.

The papers contained in this volume represent the yield of one such interdisciplinary effort. The chapters are diverse, but the aim is unified: to characterize food as a fundamental environmental influence exerting an effect on man at every level of biological organization, and to characterize man as an organism capable of reacting to food in a multiplicity of ways. The volume has been arranged to follow a progression from macrocosm to microcosm--from historical and global perspectives on food and nutrition, to analyses of the impact of specific nutrients on the individual.

The opening chapter, by Nobel Laureate Arthur Kornberg, introduces the science of nutrition--the fundamental discipline which underlies the entire volume. Historian Lynn White, jr. and geographer F. J. Simoons, examine variations in human nutrition across time and distance. E. J. Ojala and James E. Austin discuss the world food crisis as it relates to population growth and economic conditions, and consider strategies for resolving it. E. M. DeMaeyer and Cypriano A. Canosa describe the human impact of malnutrition and methods for measuring and diminishing it.

The next bloc of five chapters is given over to various aspects of human behavior and decision making relative to nutrition. J. Trémolières considers the various modes of response which man exhibits towards food, and the manner in which food may be used to gratify emotional as well as physical needs. I. de Garine examines the cultural factors which underlie the phenomena of prestige foods; William J. Darby outlines the criteria to be employed in making rational benefit-risk decisions concerning food safety. Frederick J. Stare describes the trends in food faddism, and Hans D. Cremer the educational efforts which must be undertaken in order to develop a cadre of nutritional experts capable of effecting an improvement in patterns of nutrition.

Finally, the last bloc of six chapters describes the effect of nutritional factors on the health and development of the individual. Hugo E. Aebi discusses the manner in which different proteins may be combined to achieve an optimum intake of amino acids in the diet. Jean Frézal discusses possible patterns of interaction between food and the genes controlling development. Hanuš Papoušek discusses the effect of food on the psychological development of the human infant, while John Dobbing considers the effect of malnutrition on the developing brain. Ettore Rossi discusses the relationship between nutritional factors and somatic growth, and Niels Räihä examines the specialized nutritional requirements of low birth weight infants.

Taken together, these eighteen chapters provide an interdisciplinary view of the complex web of interactions between food, man, and society. Under the shadow of the world food crisis, scholars once again find themselves gathered together to contemplate the relationship between food and those who consume it, just as the sages of ancient India some twenty-five centuries ago, watched the sacrificial fire and tried to express the mystical relationship between the food offering and the flames:

> I am food! I am food! I am food!
> I am a food-eater! I am a food-eater!
> I am a food-eater!
> I am a fame-maker! I am a fame-maker!
> I am a fame-maker!
> I am the first-born of the world-order,
> Antecedent to the gods, in the navel of immortality!
> Who gives me away, has indeed aided me!
> I, who am food, eat the eater of food!
> I have overcome the whole world!

Taittiriya Upanishad*

---

*Robert Ernest Hume, "The Thirteen Principal Upanishad," London and New York, Oxford University Press, 1921, p. 293.

NUTRITION AND SCIENCE

Arthur Kornberg, M. D.

Professor of Biochemistry
Stanford University
Stanford, California

The unifying theme of this meeting on human development is nutrition. Some of you may wonder about my credentials for being here to introduce a symposium on nutrition. With some effort I can think of a few.

First, as a medical student at the University of Rochester I took a course in nutrition which may have been unique even then in medical school curricula. It was entitled Vital Economics. It was given by John Murlin. This course provided a more systematic and rational exposure to the elements of nutrition than any medical school course we have given for the past sixteen years at Stanford.

Second, I spent the first three years of my research career studying rat nutrition in a laboratory in which I trace a direct lineage to Joseph Goldberger. I had completed an internship in internal medicine in 1942 and entered military service as a commissioned officer of the U. S. Public Health Service. I was on sea duty as a ship's doctor but was transferred with the eager concurrence of a captain exasperated by my inattention to naval etiquette. I was transferred to the National Institutes of Health (NIH) and assigned to the nutrition laboratory, which had been founded by Joseph Goldberger.

I would like to talk briefly about Joseph Goldberger, who is one of my favorites among microbe hunters and hunger fighters. He grew up in New York's lower East Side.

Despite poverty he managed to go to City College and get medical training at Bellevue Hospital Medical College, now New York University Medical School. He joined the United States Public Health Service in 1899 for adventure. For the next ten years he made important contributions to the understanding and control of several infectious diseases, including yellow fever and dengue fever, each of which nearly killed him.

In 1914 he was sent to the southern United States to find the organism pellagra. Epidemics of this disease afflicted hundreds of thousands of people each year. They had skin lesions, weakness, diarrhea, and mental derangements. Many were committed to asylums. The economic effect of the disease was widespread on the cotton plantations where the workers were afflicted.

Goldberger observed that in institutions with severe epidemics, inmates were affected, but the staff people were not. This was a remarkable disparity for a contagious disease. He noted too, that whereas inmates ate corn bread, grits, molasses, and fat back, the staff ate meat, milk, and vegetables. When he fed the inmates' diet to dogs, the dogs developed blacktongue, a canine analogue of pellagra. Pellagra patients placed on good diets were miraculously cured; and hopelessly insane people were well enough to leave the asylums.

Goldberger proved by controlled experiments that pellagra is a dietary deficiency disease. This landmark discovery of a nutritional deficiency led him to intensive assays of foods for their antipellagra value. Nutritional research was a novel departure for the Hygienic Laboratory (as the NIH was then called), a laboratory which had been oriented to infectious diseases. Goldberger's discovery of blacktongue in dogs led directly to the finding in 1937, eight years after his death, that nicotinic acid is the antipellagra vitamin.

The Nutrition Section at the NIH, which I joined in 1942, was directed by W. H. (Henry) Sebrell; my immediate adviser was Floyd S. Daft. The *L. casei* factor was close to being isolated and identified as folic acid. We tested precious samples of liver concentrates and crystalline preparations of this new vitamin, sent to us by other laboratories. We established the capacity of folic acid to correct anemia and granulocytopenia induced in rats fed sulfa drugs or a highly purified diet. These studies led to others in which we evaluated the contribution of intestinal bacteria to the vitamin needs of the rat. We also examined the nutritional requirements for blood cell formation under stressful circumstances.

I have now cited two credentials to speak on nutrition: a proper course in nutrition and publication of twenty experimental papers on the subject. My third qualification is for many people the most persuasive. For all of my adult life, some forty years, I have been curious, and, of course, concerned each day about human nutrition, my own.

In 1945, I decided to switch from nutritional work to enzymology. On the one side I had become dissatisfied with feeding rats. While it was gratifying to know that folic acid prevented and corrected blood cell deficiencies, I failed to see how I would ever learn, without biochemical approaches, _why_ it was needed for the manufacture of blood cells. On the other side I was attracted by the beat of the new drummers. I was enthralled by the "one gene-one enzyme" concept of George Beadle and Edward Tatum based on their _Neurospora_ work. An even greater revelation for me were the feats of the enzymes described in the work of Otto Warburg, Fritz Lipmann, Herman Kalckar, Carl Cori. and Severo Ochoa. Here was a window on a new world of science: enzymes of astonishing specificity and catalytic potency linked the combustion of foodstuffs to the generation of ATP which made the cell grow and the muscle move. What fantastic natural poetry!

It is thirty years now since my defection from nutritional research. I have all along retained a sentimental attachment. In the last few years I have become increasingly impressed that a renaissance of nutritional science was needed and timely. I am now concerned that it is overdue. I won't presume to suggest or evaluate, especially to an expert audience such as this, important areas or particular problems in nutrition. I simply am not well-enough informed. What I can attempt, as a sympathetic outsider, is to look in on the discipline of nutrition as a whole and to express my feelings about what I perceive to be its strengths and weaknesses.

Most of you would agree that nutrition enjoyed a "golden age" in the first half of this century; in the United States, nutrition dominated biochemistry. For a decade or so before World War II, emphasis in Europe began to shift to enzymology and intermediary metabolism. After the War, this shift became pronounced in the United States as well. In the past two decades macromolecules: their structure, biosynthesis, and regulation - became the dominant fashion. There was an increasing influence of genetics on biochemical research. But even under this strong genetic influence, _Drosophila_ genetics, despite its advanced state, was submerged in the wake of a rush to bacterial and bacteriophage genetics.

Fashions in science change every few years. They have again changed dramatically in the last five years. For example, _Drosophila_ flies again. There is a reawakening of interest in the development of cells, organs, and organisms. In this latest shift back from molecular events to cellular phenomena, it becomes clear once again how dominant a force nutrition is in biology. There is every reason now for an invigoration of nutritional science to a status equal with genetics and biochemistry.

But there is an ominous cloud on the horizon which threatens any burgeoning of nutritional science. It is the confusion between nutrition as a natural science and nutrition as the art of feeding people. It is a confusion analogous to that between medical science and health care. One sees everywhere efforts to broaden the scope of nutritional science far beyond the range of its scientific disciplines. In overreaching its scientific domain, nutrition, like so many other human endeavors, becomes prey to prejudice, chicanery, and other perils of our social culture.

I am now venturing on ground that is heavily mined. It is strewn with the remains of those foolish or intrepid enough to join this controversy. But I won't retreat. We must be aware that when nutrition tries to encompass medicine, agriculture, economics, psychology, and anthropology as well as biology and chemistry, it becomes more a social and political activity and less a science. The controversy boiled over this year in response to a report, known as the Neuberger Report on Food and Nutrition Research, which was issued jointly by the British Agricultural and Medical Research Councils.

This report emphasized that "human nutrition is mainly concerned with defining the optimum amounts of the constituents of food necessary to achieve or maintain health." It recommended more research effort in the biochemical aspects of nutrition. It also deplored the relative neglect of human nutrition and urged more clinical and epidemiological studies. Such a recommendation, made by a large committee drawn from diverse areas of nutrition, seems to be very broad and balanced.

Not so to others. John Yudkin inveighed against it because of what he regarded as excessively narrow focus. There is, in his view, inadequate concern in the Report with economics, anthropology, sociology, demography, and psychology. In his words: "Nutrition is the one science that can least afford to remain in the laboratory; it concerns every single human being, every single day of his life." T. B. Morgan expressed bitterness about the Neuberger Report because it failed to stress "the inseparable interrelationships between agriculture and nutrition." He deplored the lack of concern with training graduates from food science, nutrition, catering, and dietetics for research in <u>social</u> nutrition.

In joining this argument, I would like to make some of my feelings clear at the outset. I believe that an optimal balance must be sought between basic research in science and its application to the needs of society. Usually it is prudent not to mix the two. Generally, individuals have a talent or inclination to do one or the other. This may even be true of institutions and nations. Unfortunately, the imperfections of our national and world societies impose a heavy penalty on a group or nation which contributes more than its share of basic knowledge but fails to exploit its technologic applications. Britain and Israel are examples of this extreme; Japan, by contrast, has prospered by honing scientific advances made elsewhere for developing competitively superior commercial products.

Before commenting on the scope and practice of nutrition, in which I have no professional experience, I would like to comment about medicine, in the organization and practice of which I have had some exposure. I find the parallels between the two fields instructive.

In the debates about medicine and health care in the United States, arguments usually center on two charges. One is the failure to apply new knowledge to medical practice. This is really a minor issue. There have been some instances of long lag periods in the appreciation and implementation of basic findings for clinical use, but these lags have been due more to a lack of imagination than indifference. Yet the damage done by these failures in initiative was not nearly as great as that from premature and overzealous applications of new findings to clinical use.

The second charge is the inadequate and inequitable distribution of medical care. This is a serious issue. For example, we have the means to eradicate venereal disease and malaria and yet they flourish in the world today. This is an exceedingly complicated issue bedeviled by economic, social, and geographic factors. But at least as serious as the failures in the underdistribution of medical care is its overdistribution. There is a threat that the profligate use of prohibitively expensive diagnostic and therapeutic procedures will be sufficient to bankrupt us in the United States. As one extreme example, the bill for 80,000 coronary arteriograms a day recommended by one authority would come to 10 billion dollars a year; the bypass surgery which would be indicated by such radiologic diagnoses would cost 100 billion dollars a year. Absurd? Perhaps. Yet there is another example which already wastes billions of dollars not to mention thousands of lives, each year. It is the performance of at least twice as much surgery in the United States as competent authorities deem necessary.

There is a third issue which I regard as the most important of all. It astonishes and dismays me that in all the legislative, and even academic, discussions of medicine and health care, it is given so little attention. It is the inadequacy of available knowledge for prevention and cure of disease. Applying what we don't know won't help a single patient. If cost-benefit analyses were calculated for the development of vaccines, antibodies, and hormones, the benefits in dollar savings would prove to be enormous compared to the investment cost, even without measuring the values of enhanced well-being.

We quickly forget the bygone scourges of polio, pneumonia, and untreated diabetes. It was basic research in virology, immunology, and cell culture that gave us the polio vaccine. Otherwise we would be spending billions on more elegant iron lungs and Sister Kenny physiotherapy centers. It was basic research on understanding the lysis of bacterial cell walls that revealed penicillin. Can you imagine medicine today without antibiotics?

I find the analogies between medicine and nutrition compelling. In nutrition, as in medicine, similar charges are made about failures in clinical application of basic knowledge and the maldistribution of nutritional knowledge. Unfortunately, there are tragic deficiencies in the use of nutritional knowledge for human welfare. An example cited in the Neuberger Report is that at least 10,000 children go blind in India each year from a deficiency of Vitamin A which could be prevented by five cents worth of the synthetic vitamin. Thomas H. Jukes has pointed out that the needed allowances for ten vitamins and eight essential minerals could be supplied at a bulk cost of less than one dollar per person per year. But effective distribution of vitamins and minerals is far from being achieved, for socioeconomic, political, and logistic reasons. This is a deplorable state of affairs and should arouse everyone's indignation. Yet this is a failure for which our society is to blame, not the nutritional scientist.

Where the responsibility does rest with the nutritional scientist is in the inadequacy of available knowledge. We need to know more about the nutritional requirements for optimal function in any individual under normal and stressful circumstances. We do not have a sufficient understanding of the diversities among individuals in a given culture, and the diversities among the many cultures in our world society. There are serious questions about proteins, cholesterol, sucrose, and lipids.

In nutritional science as in other science we focus on the problems that face us and take for granted the enormous progress behind us. That is as it should be. Yet we cannot expect the debate about vitamin C and vitamin E to be settled in the newspapers. Management of obesity must not be left to best-selling charlatans. The serious questions in nutrition will be resolved only in the laboratory with experimental animals and in the field with properly controlled clinical trials. I don't underestimate the extraordinary difficulty of these experiments, but there is no other way.

I want at this point to reflect on the history of science for the important guidance it can give us in planning the future. I was struck in reading a recent article by Gerald Holton, a science historian, by a quotation from the outstanding physicist, P. A. M. Dirac:

> The point of view of the historian of science, Dirac said, is really very different . . . from that of the research physicist. The research physicist, if he has made a discovery, is then concerned with standing on the new vantage point which he has gained and surveying the field in front of him. His question is, Where do we go from here? What are the applications of this new discovery? How far will it go in elucidating the problems which are still before us? What will be the prime problems now facing us?
>
> He wants rather to forget the way by which he attained this discovery. He proceeded along a tortuous path, followed various false trails, and he doesn't want to think of these. He feels perhaps a bit ashamed, disgusted with himself, that he took so long. He says to himself, What a lot of time I wasted following this particular track when I should have seen at once that it will lead nowhere. When a discovery has been made, it usually seems so obvious that one is surprised that no one had thought of it previously. With that point of view, one doesn't want to remember all the work that led to the making of the discovery.

We see in these intimate revelations of Dirac, the theoretical physicist, how remarkably identical his attitudes and feelings are to those of experimental scientists. We must always reflect on our own experiences and on the history of science to keep the essence of this culture clearly in mind. Otherwise we will be engulfed by the pressures and confusions of everyday life and contribute nothing of lasting value.

It was only a few centuries ago that a few men disciplined themselves to ask small and humble questions. This was the crucial element in the development of our scientific culture. As Victor Weisskopf states it:

> Instead of reaching for the whole truth, people began to examine definable and clearly separable phenomena. They asked not what is matter but how does a piece of matter fall or how does water flow in a tube; not what is life but how does blood flow in the blood vessels; not how was the world created but how do the planets move in the sky. In other words, general questions were shunned in favor of limited ones which were to some extent more fully answerable.
>
> In time this restraint was rewarded as the answers to the limited questions became more and more general. The renunciation of immediate contact with absolute truth, the endless detour through the diversity of experience, allowed the methods of science to become more penetrating and their insights to become more fundamental. The study of moving bodies led to celestial mechanics and an understanding of the universality of the gravitational law. The study of friction and of gases led to the general laws of thermodynamics. The study of the twists of frog muscles and of voltaic cells led to laws of electricity that were found to be the basis of the structure of matter. Einstein considered this development to be the great miracle of science; in his words, "the most incomprehensible fact of nature is the fact that nature is comprehensible." By means of this detailed questioning, man has created a

> framework for understanding the natural
> world, a scientific world view.

At this symposium we have experts from the whole spectrum of nutrition, from nutritional science to ethnology, from agricultural science to geography, from rat nutrition to human nutrition. This is a remarkable opportunity to enhance one's appreciation of the breadth and encompassing character of nutrition, an opportunity to be apprised of the problems and progress in the diverse disciplines that constitute nutrition. For the nutritional scientist, this symposium will enlarge his capacity to treat his subject more humanistically. It may even facilitate the application of scientific nutrition to help relieve the world's food problems. Without this input the food crisis will surely get worse. Yet the nutritional scientist must struggle to keep his focus on science.

Realistically, an experimental scientist can hope to apply his knowledge by interacting with a closely neighboring discipline. In my own case, I have been fascinated by my experience as a scientific adviser to a small and enterprising company whose mission is to provide more accurate and rational drug therapy through the use of polymer science. As one outgrowth of this approach, there has been a highly promising development for the food industry. This is the synthesis of polymers with covalently attached color dyes, or sweeteners, or antioxidants, or other food additives. These polymers pass through the intestinal tract without absorption. In some instances their biological activity far exceeds the values anticipated from the content of the low-molecular-weight dyes, sweeteners or antioxidants attached to the polymers. They promise to be excellent vehicles for enhancing the appearance, taste, and stability of foods and for avoiding the possible dangers of various food additives.

In concluding, I want to express my gratitude for this opportunity to be with you and the stimulus to think about nutrition. At the beginning of my research career, I was devoted to nutrition because I enjoyed it and saw it as the bridge between biochemistry and physiology. Having been on the biochemistry side for the past thirty years, I see the bridge from the same vantage point. Nutrition is still a major passageway between biochemistry and physiology. Of course, it provides access to many other areas, such as clinical studies, agriculture, and the proper feeding of populations. Without in any way denigrating the importance of the many disciplines allied to nutrition, I still emphasize that, in the long term, the rate-limiting factor in progress in nutrition will be the fund of knowledge available from the experimental science of nutrition. Nutritional science, like any other, builds its structure a brick at a time. It is only by this humble, undramatic brick-laying that the great bridges and edifices of science are built.

FOOD AND HISTORY

Lynn White, jr., Ph.D.

University Professor of History, Emeritus
University of California
405 Hilgar Avenue
Los Angeles, California 90024

Brief discussion of a topic like "food and history" demands clear statement of what one intends to do. The two chief temptations are either to present part as though it adequately mirrors the whole, or else to schematize the whole in such a way as to lose contact with concrete experience. In this paper I have chosen to commit both sins in sequence, hoping that they may cancel each other sufficiently to give us some glimpse of reality.

First, simply to illustrate the specific complexities and puzzles facing any historian of the relation of food to its wider context, I shall speak of proteins in medieval northern Europe. Then I shall invite you to go with me into high orbit trying to envisage the global pattern of the history of food and its effects on human activity.

PROTEINS IN MEDIEVAL NORTHERN EUROPE

Our understanding of the connection between cultural dynamics and protein foods, whether animal or vegetable, is still rudimentary. In the Western world, for historical reasons that I shall indicate, we have a bias favorable towards meat and dairy products, and this has almost certainly warped some of our supposedly objective scientific findings on the subject. Nevertheless, the weight of the present evidence indicates that diets deficient not merely in calories but specifically in amino acids increase listlessness and vulnerability to disease. What is a bit less firm but even more ominous for the increasingly hungry parts of our world is indication that insufficient protein in a pregnant mother's food handicaps the development of her fetus, and that continued dearth of proteins may permanently retard a child's mental development. To what extent early disasters may later be recouped by remedial diets remains in doubt. No historian would risk the hypothesis that the abounding vitality of Chaucer's or Shakespeare's England was based primarily on a more frequent and widespread munching of mutton chops. Yet it would be equally rash to assume that the incidence of mutton chops is unrelated to the merriness of those Protean generations.

In the Graeco-Roman world herding and farming seem to have remained, in general, separate activities. Oxen, of course, were needed to plow the fields, and pigs, chickens, and geese scavenged around the villages. But sheep, goats, and cattle were normally pastured on domains separate from the cultivated areas. After the Germanic invasions this pattern began to change.

Not in the Mediterranean area but north of the Alps and the Loire (1), a new and more efficient type of plow was introduced which broke the soil so violently that cross-plowing was unnecessary and was replaced by harrowing. Cross-plowing had produced squarish fields that were normally hedged or walled. The new plow worked best producing long strips in big open fields unbroken by barriers except at their furthest reaches. From the beginnings of agriculture farmers who had gone beyond the slash-and-burn system had recognized that to maintain its fertility a field must be left fallow at least on alternate years. The result, in northern Europe, was a two-field system, with half the arable uncultivated in a given season. When the new plow came into use, the great open fields apparently led to more frequent combination of herding with cereal farming, because sheep and cattle could as easily be set to graze on the wide sweeps of stubble and fallow as upon wild pasture. When this was done their droppings were not wasted but fertilized the crops of the next planting. Manure became so valuable a commodity that herds were increased and farmers began to reap hay systematically to stall-feed their animals during the worst of the winter, collecting the resulting fertilizer. The haying instrument was the scythe. Roman scythes have been found, but they are rare. By the ninth century, however, haying was so basic to northern European agriculture that Charlemagne proposed to rename July "Haying Month," and an illustrated calendar prior to 830 personifies July as holding a scythe, whereas August, the "Harvest Month," carries a sickle. In other words the system of "mixed" farming - combining cereals with production of meat, butter, and cheese - which is widely regarded as distinctive of northern Europe and much of North America today, had its origins in Frankish times. It implies a diet fairly high in proteins.

The peasants, however, moved onward in their innovations. With rare exceptions, the two-field system involved autumn plantings and late spring or early summer harvests, and this remained the Mediterranean pattern. North of the Alps and the Loire, however, summer rains made possible a spring planting with harvest in the later autumn. To have two plantings and two harvests each year offered great advantages. The chance of famine was reduced by the probability that both crops would not fail. The labor of plowing was more evenly distributed over the year. Shortly before 800 we begin to see northern peasant communities operating a three-field rather than a two-field rotation: one fallow, one planted in the autumn, and one, in the spring. It was found that some crops did better in one planting than in the other: in general, wheat and rye were sown in the autumn, while oats, barley, and legumes were planted in the spring. An English game-song gives us a picture of the summer fields:

> "Do you, do I, does anyone know
> How oats, peas, beans and barley grow?"

Not only protein rich peas and broadbeans, but also lentils and chickpeas were common field crops of the spring planting, and since the bacteria on their roots fixed nitrogen, they helped to sustain the fertility of the soil under the more intensive rotation. In areas of summer rains where the soil was rich, transition from two- to three-field rotation, therefore, increased the production of vegetable proteins in relation to carbohydrates. That in the minds of contemporaries these two types of crops ranked equally is indicated by the lament of a Norman chronicler that a drought in the summer of 1094 had seared both "the grain and the pulse" in the fields.

For intricate reasons, in villages where waste land could be reclaimed for arable fields, the new rotation enabled a peasant to put under crops about fifty percent more hectares annually than had been possible in the older rotation. Moreover, in many areas of three-field rotation it soon became possible to substitute the horse for the ox as the normal draft animal, from the Channel to the Dnieper. By about 800 a new system of harness reached Europe from Asia that increased the pulling power of a horse between four and five times. By about 900 the nailed horseshoe came into use, greatly improving the staying power of draft horses. Finally, the new three-field rotation raised production of oats, the best feed for horses. Mediterranean peasants, lacking summer rains, were bound to two-field rotation. Since they could not produce surplus oats to feed many horses, they continued to use the ox as their draft animal. Horses, however, can work longer than oxen in a day, and while they pull no harder than do oxen, they draw the plow more rapidly. In the Slavic East, a horse was judged to do twice the plowing of an ox. If, as has been mentioned, changing from the old to the new rotation increased the potential arable of a village each year from say, 1000 hectares to 15 hectares, it follows that, if the peasants were also able to substitute horses for oxen as plow animals, their potential was 3000 hectares, an increase of 200 percent over the two-field system. Northern medieval agriculture became vastly more productive than it had been, while becoming less labor-intensive and more land-intensive and energy-intensive. Since, as we have seen, the planting of legumes increased, under the new system, even more rapidly than that of grain, we may safely guess that production of pulses rose by 250 to 300 percent in the more favored areas.

My colleague at UCLA, Jay Frierman, a sharp-eyed historian of ceramics, permits me to cite an unpublished finding that coincides with this conclusion. In the early Middle Ages the most common form of cooking pot is small and cylindrical. From strata of the later ninth and early tenth centuries in the north, we begin to excavate considerably larger, globular pots that to American eyes are immediately recognizable as beanpots designed for simmering legumes over a slow fire. Clearly, in those regions at that time the pulses were becoming more available and were eaten in greater quantities than formerly. Once more a game-song:

"Pease porridge hot; pease porridge cold;
Pease porridge in the pot
Nine days old."

One suspects that, in Europe, these pots usually held not only legumes but also - as they do in modern America - some bits of pork. It is clear that most of the meat eaten by the medieval lower classes was pork. A few bulls were kept for breeding; most male calves were castrated and saved for plowing in two-field regions; in three-field regions they probably ended up as veal on upper-class tables. Cows provided butter and cheese. Sheep were primarily suppliers of wool rather than meat. Hens and geese were valued as egg producers. Only old oxen, sheep, cows, geese, and hens would be eaten, along with an occasional gander or cock. The pig was different. Like the chicken, it could scavenge and eat almost anything, although it fared best on the acorns and beech mast of the remaining forests. But, save for its hide, it supplied nothing but meat. Each year a sow produced two litters of about seven piglets each, and these grew quickly. Until the late Middle Ages, every calendrical picture of butchering that I have seen shows the slaughter of a pig.

The sardonic Edward Gibbon confronted the intellectuals of his day with the horrid speculation that if Charles Martel had not repulsed the Muslim invaders so decisively at Poitiers, "the interpretation of the Koran would now be taught in the schools of Oxford, and her pulpits might demonstrate to a circumcized people the sanctity and truth of the revelation of Mahomet." To this present audience may I simply suggest that if Spanish Muslim forces had overwhelmed the Franks and converted Europe to Islam with its absolute aversion to pork, the result might have been - for the little people at least - nutritionally damaging?

Yet, by curious coincidence, at the moment of Martel's victory, the Christian Church was beginning to enforce upon the newly baptized Germans a prohibition of eating horse flesh. In his remarkable study of food tabus in the Old World, Frederick Simoons has shown convincingly that this ban had no basis in utility, sentiment, or health: it was purely a religious matter (2). From India to Iceland many great Indo-European gods were worshipped with horse sacrifices, after which the faithful feasted on the offering.

When St. Boniface and his missionary colleagues brought the Germans into the new faith, abstaining from horse meat seems to have been one way of showing conversion. It is said that as late as 1629 in France a groom was executed for eating it (3). In practice the ban extended even to mare's milk. I have found no hint of its use in medieval Europe even though, in the form of kumis, it provided the favorite beverage of the nomadic cultures of Central Asia - in which, incidentally, horses were eaten quite customarily, as camels are among Arabs. I suspect, however, that even if the Church had refrained from banning horsemeat, it would have made little difference in the protein diet of most people. Horses are large animals that reproduce slowly. If they had been eaten, most of their meat would have been consumed by the aristocracy, as the beef evidently was. Smaller, quickly reproducing animals like pigs and chickens were what the common people needed.

The rabbit was, in effect, a medieval addition to this repertory (4, 5). Unlike the hare, which does not breed easily in captivity, the rabbit does. By the first century B.C. (perhaps as a rival of the domesticated dormouse) it had been introduced from its native Spain to Italy, but it was evidently not widely exploited until after the Frankish era: Charlemagne's _Capitularium de villis_, that is so detailed about the products of the imperial manors, does not mention the rabbit. Sometimes the rabbit was kept in warrens by aristocrats; sometimes, simply in hutches by peasants.[1] Its first appearance in England is in 1186, and evidently it was then a newcomer. In 1341 a chronicler of Milan tells us that the city and surrounding villages were filled with rabbits. To peasants they were attractive because they ate grass and vegetable waste, they provided substantial meat, and their fur had value. Moreover they are so prolific that at Easter they fuse with a more ancient symbol of fertility and lay eggs.

Undoubtedly throughout the Middle Ages selective breeding was improving the useful properties of both plant

---

[1]Domestication of the rabbit lead to the "invention" of the ferret, a form of polecat not found in nature, to assist warren keepers from flushing rabbits from their h

and animals, including their ability to supply protein food; but of this we know remarkably little as yet. Let me offer simply one piece of evidence that the size of cattle was gradually increasing. The water buffalo was introduced to Europe from South Asia probably by the end of the sixth century. It was well established in Italy by the middle of the twelfth century. Its meat was not relished, but its rich milk was used to make mozzarella. Contemporaries considered buffaloes two or three times as strong as oxen, and several of the thirteenth century encyclopedists mention that to govern them a metal ring was placed in their nostrils (6). No such rings are mentioned in descriptions of oxen or bulls. I cannot discover when European bulls first received noserings, presumably by transfer from water buffaloes. When that date is discovered we may be confident that the size of at least certain varieties of cattle and consequently available protein, had been increased by breeding.

What of fish as protein food in medieval Europe? The changing factor as regards fresh water fish was the rapid diffusion of the watermill. This earliest of the power engines was invented in the first century B.C., but spread only slowly during Roman times. In the Germanic kingdoms it became widely used, and from the ninth century began to be employed increasingly for industrial processes other than grinding grain (7). Reliable statistical material is rare from the Middle Ages, but the Domesday Book of 1086 lists 5,624 mills in about 3000 English settlements. Since the more efficient overshot wheel was preferred, most mills, at least on small streams, had millponds. That these became fishponds as well is indicated by the frequency with which feudal dues on mills were paid in eels and fish (8). The Church's rules for fasting on Fridays, during Lent and at other times, put a premium on securing fish, and all the evidence indicates that even in inland regions considerable quantities were available, although, as usual in greater measure to the rich than to the poor (9).

As regards salt-water fish, a technological breakthrough, probably of the later fourteenth century, added considerably to Europe's supply of cheap protein. Many fish, like cod, had little fat and could be preserved by the ancient methods of drying, salting or smoking. Herring were a different matter. Herring are replete with

unsaturated fat that quickly goes rancid on exposure to air. Moreover herring run in immense schools: catches were often so large that even a quick sail to harbor and fast handling did not suffice to preserve most of what had been caught. Legend says that a Netherlandish wholesale fish merchant, Willem Beukel of Biervliet, who has been dated anywhere from the late thirteenth to the middle fifteenth centuries, solved the problem. He sent fishing boats into the North Sea equipped with salt and good barrels. As soon as they were caught, herring were carefully gutted, salted, and laid compactly, head to tail, in airtight barrels that prevented oxidation. In northern climates such fish would last indefinitely; even in Italy, so it is said, it was good for a year. Archival search has discovered a fish merchant William Beukel living from 1388 to 1396 at Hugevliet not far from Biervliet (10). Whether he, in fact, invented the new method of preserving herring or whether he is only a symbol of a revolution in fish-processing then taking place, we may never know. But the fact is not in doubt that in this period the production of cheap herring packed in barrels grew enormously, that it was exported widely, and that most of it was eaten by the lower orders.

Let me add a final problem in the medieval history of proteins. As we have seen, the agricultural revolution of the ninth to eleventh centuries in northern Europe brought a great increase in the productivity of peasant labor. The new supplies of food were at least one of the reasons for the startling upswing of population from the middle of the tenth century onward to the later thirteenth. However, Josiah C. Russell, the leading American demographer of the Middle Ages, noticed that northern aristocratic families in that period were producing an even larger proportion of children than plebeian families. The result was high social tension. Younger sons, unable to find support on increasingly scarce fiefs, either entered monasteries like Cluny (11) or else went adventuring as mercenaries: the beginnings of the Reconquista in Iberia, the Norman conquests of both England and southern Italy and Sicily, the German _Drang nach Osten_ as well as the earlier Crusades, must be understood as, in part, the result of a sudden surplus of young men whose self-image was that of the warrior. Russell asked why feudal wives were so newly prolific. Evidence has been rapidly accumulating to show that nursing a child reduces a woman's ovulation: the effect is maximal for nine months and considerable for about fifteen months (12). From what little could be learned of the spacing of their children, it is clear that some of the great ladies of the time were not suckling their own babies but were turning them over to wetnurses, thus becoming pregnant again more quickly than would otherwise have been the case. Russell proposed the hypothesis of a "nursing revolution" among the upper classes of northern Europe in that era, involving a change of fashion which wasted a small, but socially significant, proportion of that miraculous protein food, human milk (13). To date it remains nothing but hypothesis: such change is the sort of thing that preachers and satirical poets might well notice and denounce, but thus far no documentation has been found. Yet the observed fact of feudal fecundity demands explanation, and until a better one is offered, we must treat Russell's with respect.

What does our survey of northern medieval protein production and consumption add up to? Fernand Braudel, who has done so much to broaden historical perspectives on the later Middle Ages and Renaissance, has recently insisted that, from about 1350 to about 1550, what he calls (p. 142) "l'Europe carnivore" was the most meat-eating culture in the world, and the lower classes shared this dietary pattern (14). The modern notion, he says, that medieval peasants and workmen ate wretchedly arises from the fact that from about 1550 to 1850 meat consumption by such groups, in fact, declined considerably, and the modern dogma of progress has led to the assumption that in earlier times they, therefore, must have eaten worse.

Dietetically, of course, the issue is proteins, not meat: a meal of wheat bread and peas can supply as nourishing a mix of amino acids as one of beefsteak. But meat tastes better than peas, at least to Western palates. For some three centuries, ca. 950-1250, after the innovations of the three-field system and the horse-drawn plow came into play on food production, there would seem normally to have been enough food, including proteins, in northern Europe for a rapidly expanding society. About 1050 a German cleric wrote a picaresque Latin poem, "Ruodlieb," in the course of which he describes a village where his hero spends a night (15). The peasants are by no means poverty-stricken: the village consists of big houses built around courts, including barns, stables, and storehouses. They have many cattle, horses, sheep, goats, hogs, chickens, geese, and bees. The people eat plenty of food, including meat. Yet the author is not gilding peasant life: manners are rough and the older men seem particularly dirty. What he gives us is an image of eleventh century South German peasant life in the flush of the new agricultural prosperity, before the growth of population began to undermine lower class well being.

Few social historians of the Middle Ages now doubt that by the later thirteenth century the period of expansion was drawing to a close. Population growth was outrunning increase of the food supply; indeed, infertile soils that should never have been put under cultivation were relapsing into the wild. Hunger and disease grew. Then came, from 1347 to 1349, the Black Death, with recurrent epidemics every few years. It is generally estimated that by 1400 Europe's population had been reduced to half what it was in 1347.

How did this affect proteins? Plagues destroy people, not property. The survivors in the fourteenth century were heirs of the dead, and we may assume that many, in their new affluence, gave up working with their hands. Thus the labor force was reduced somewhat more than the total population. Labor became a scarce commodity and - despite restrictive legislation - wages rose. For a few generations the poor became less poor, and they began to spend part of their new income on meat. The first calendrical picture known to me of the slaughtering of a steer, as distinct from a pig, is of the fifteenth century from the middle Rhine (16). When an established iconography is altered, something significant is happening in society. Moreover the drastic depopulation of Europe meant that much arable land became pasture for cattle and sheep simply because there were too few hands to cultivate it. Thus the new market for meat found its supply. It is for such reasons that I believe that Braudel is correct in maintaining that about 1350 Europe entered a new era of carnivorousness. I suspect, however, that, from the early Middle Ages, the peoples of northern Europe at least had enjoyed, with fluctuations, a diet higher in proteins than the people of any other major society.[1]

THE GLOBAL PATTERN OF THE HISTORY OF FOOD

I have skimmed in detail the history of a single nutritional element, protein, in one small part of the world, northern Europe, during a brief span of human experience, roughly 700 to 1500 A.D. My purpose has been to indicate the variety of forces, circumstances, and cultural preferences that must be pondered in any study of food and history.

Now permit me, even more briefly, to approach the relation of food to history from the other end, that is, in its largest pattern.

The first great event in the history of human food was presumably the invention of cooking. Charred bones at Chuo Kuo Tien have led to the assumption that Peking man roasted his meat. Some think the evidence inconclusive: he may simply have been burning refuse bones. I find it hard to believe that people have had fires for warmth and protection without quickly discovering the benign effect of fire upon food. All kinds of meat can be eaten raw, but many kinds of vegetable matter now regarded as food are inedible unless cooked. I have not found a detailed study of the way in which the taming of fire widened the options for human nutrition, but clearly it gave man a great competitive advantage. Since in most very primitive societies of the recent era men did the hunting and women the gathering, it is safe to assume that the use of fire both increased the responsibilities of women and improved their status somewhat.

---

[1]In Oceania the numberous groups of cannibals disliked the rank flavor of Westerners (as Westerners decline to eat the meat of predators) and much preferred Chinese and Malays who had a diet higher in carbohydrates.

Scholarship in Western societies has only recently begun to realize, and to escape from, its own profound historical conditioning. Our fascination with the Graeco-Roman world, and especially our reverence for the Hebrew scriptures, have motivated extraordinary adventures in Mediterranean and Near Eastern archaeology, reaching into upper Egypt, Mesopotamia, and Iran. Since in these regions more has been dug, more has been found than elsewhere. Myths are seldom eradicated even after they are rejected. Cain was the original herdsman and Abel, the primal plowman, so it was inevitable that even archaeologists who had forgotten the first murder should find the beginnings of both animal and plant domestication to be Near Eastern.

In very recent years the explosion of radiocarbon dating, now expanded by the collagen method, has combined with the general cultural shock suffered by the Occident both to discredit our traditional ethnocentrism and to question the methods by which we have drawn conclusions about domestication from archaeological evidence. A major early symptom of this development was the methodological criticism of Higgs and Jarman (17). At the present time the early history of food is taking on an entirely new shape which is constantly being reshaped by unexpected findings.

We now recognize three major centers of domestication, each with a distinctive pattern: Southeastern Asia, Southwestern Asia, and the Mexico-Peru area. At this moment it appears that the first of these was the earliest, but discoveries and reevaluations are emerging so rapidly that it would be rash to start awarding medals for priority (18).

Ants domesticated both plants and fellow insects aeons ago. For half a million years at least, human beings have existed, not of our variety in some cases but, nevertheless, human enough to have invented domestication. Yet, if we can credit our present evidence, all three of the major centers of domestication began to tame and exploit animals and plants within the time-span of roughly 8,000 to 3,000 B.C.: one percent of the minimal estimate of human existence. The mathematics of chance thus rules out independent invention of domestication in three discrete centers. We are left with the assumption of a single invention that - despite great distances and barriers -

diffused the idea of domestication even though in the early stages not many domesticates got from one center to another.[1]

From this assumption follows another: that once the idea of domestication spread, an amazing amount of psychic energy was expended on it. Since there was nothing in the Neolithic that can be called biological science, there were no Norman Borlaugs; but there must have been thousands of Luther Burbanks: keen and patient, if sometimes floundering, empiricists, always looking for promising variants and breeding them carefully. When we are able to trace the wild ancestors of domesticated plants we often stand amazed at the faith of those who bothered to fuss with them. The stubby, tiny corncobs from Mexican strata of the sixth millennium, the bitter green berries in Iran that became apples, seem to hold no promise. And what genius, in the black abyss of Brazil's prehistory, first had the imagination and daring to remove the deadly poison from manioc and then bake and eat the stuff that today feeds millions in the tropics? We shall never know anything in detail about such individual accomplishments, but we should recognize that they occurred and that we are their beneficiaries. It is well to remind ourselves that since women did much planting and reaping, these forgotten heros were doubtless as often women as men.

Agriculture produced an increasing amount of food. Population grew, and the new surplus likewise led to developed priesthoods, bureaucratic and warrior groups, merchants, and specialized craftsmen. Sometimes these clustered in cities; sometimes ceremonial centers served large but less dense communities. Psychic energy was invested increasingly in activities other than domestication and breeding. The great majority of the animals and

---

[1]The hypothesis that the end of the latest great glaciation produced simultaneously conditions favorable to agriculture in three separate regions cannot at present be sustained. We have no evidence that the climates of Southeast Asia, the Near East, and Middle America were greatly affected by the retreat of the ice.

plants now used by us were already being exploited by the early second millennium B.C. at the latest. Because orchards take some years before they bear, and because for good results complex methods of grafting and cloning are often needed, fruit was the last major wave of domestication (19). Naturally all cultivated species have continued to produce variants. I can count, however, only four major food crops that have been domesticated during the last 2000 years; oats and rye in Northern Europe during the earlier Middle Ages; buckwheat in Central Asia during the fourteenth century; and coffee in Ethiopia or Yemen soon thereafter. I know of no significant animal domestication since that of the rabbit just two millennia ago.

This should give us a pause. Can it be that we have exhausted the entire potential of nature for purposes of domestication? I suspect that the question carries its own answer. "Civilization" means "city life" etymologically. The agricultural developments of the Neolithic and Bronze Ages made city life possible, and it proved so fascinating that the better brains developed contempt for rustics and their doings. Basic new domestication practically ceased. Progressive agriculture became simply better methods of raising better strains of what already had been raised for millennia. Even the "Green Revolution" is no more than this. It is time to ponder the utility and interest of producing crops and utilizing animals that have never before been cultivated or reared. Our agronomists and animal husbandmen must recover the Neolithic sense of adventure.

With the decline of the impulse to domesticate, the history of food becomes that of the diffusion of plants and animals already tamed, primarily in the three major areas of growth.

The Islamic conquests of the seventh century opened a new chapter in Old World agriculture (2). The domestications of Southeast Asia, including India, were naturally adapted to a hot and moist climate. When Muslims penetrated India, first as merchants and travellers, they were delighted not only by citrus fruits, eggplants, watermelons, and bananas but also by such exotics as rice and sugarcane which had, indeed, been imported since Greek times to lands lying westward but had been grown there rarely if at all.

By selective breeding, Arabic-speaking agronomists produced strains of the major India crops that could thrive in dryer and cooler climates. Likewise they perfected, on the basis of South Asian models, kinds of intensive irrigation to assist such crops. As a result, by the tenth century rice and sugarcane as well as many other novel items were grown widely in the Near East and westward as far as Spain and Morrocco. Because of a cooling of northern Europe's climate in the fourteenth century, these diffusions had no effect on the trans-Alpine regions; but by the late fifteenth century rice was being grown on the plains both of Tuscany and of Lombardy. With the discovery of the Atlantic islands, sugarcane was planted there so successfully that its competition had wrecked the Mediterranean sugar industry even before the New World was discovered. Thanks largely to Islamic cosmopolitanism, by the time Columbus watched the gray Azores drop below his eastern horizon, the agriculture and animal husbandry of the Old World has been remarkably well homogenized within obvious regional limits of climates and soils.[1,2]

Columbus went searching for the costly spices of India. He returned to Spain bearing the hot Capsicum peppers that could be grown in many temperate as well as tropical cottage gardens to give zest - and vitamins as well - to the meals of even the poorest: the cuisines of Hungary, India, Szechuan, and Korea are deeply permeated by this first gift of the New World to the Old.

The European conquerors and settlers in America normally wanted familiar food; so within a few decades practically all the Old World domesticants were established

---

[1] A notable exception is the failure of the soybean to diffuse out of East Asia. The reason may be that the soybean in its home territory is largely consumed as bean curd and soy sauce, and other cultures had no precedents for such processing.

[2] Dr. Watson (2) will shortly publish a book on New Crops in the Early-Islamic World: A Study in Diffusion.

in the New. However, within a century a major part of the plants - the turkey was the only significant animal - domesticated by American Indians were similarly found not only in Eurasia but in Africa as well (21, 22, 23). There were cocoa, papayas, guavas, pineapples, avocados, and peanuts, as well as pumpkins and a great variety of squashes. There was an infinity of new kinds of beans, rich in proteins and oil. But the amazing thing about the New World carbohydrates was their productiveness. Manioc gave a vast amount of food per hectare of land so poor that no other crop would grow on it. Maize, per hectare, outproduced all the other grains. In the cooler climates, potatoes offered more calories per hectare than any other crop, and in warmer regions sweet potatoes did almost as well.

Yet while they soon became generally known, these New World crops were often slowly accepted. Cultural tastes, methods of processing and cooking, habits of cultivation, even folk beliefs long blocked their wide adoption (24): when my father was growing up in East Tennessee in the last quarter of the nineteenth century, tomatoes were known as ornamental plants but were still considered poisonous. Nevertheless, by the later seventeenth and eighteenth centuries the American domestications were being assimilated with increasing speed to global farming and culinary patterns: polenta and mamaliga were the basic peasant food of Lombardy and Romania respectively, and white potatoes were Irish. Just as the Old World's food resources had become homogenized by 1492, so by about 1800 the Neolithic and Bronze Age domestications of every part of the world had been shared with all regions where they were viable. This is obviously not the sole reason for the population explosion of the past two centuries, but it is a major factor in it. By feedback, this uncontrolled surge of global population has made food a topic of intense concern in our own time.

I cannot undertake to summarize the history of food or the relation between food and history in the last century and three quarters. When, in 1810, Nicolas Appert published his _Le livre de tous les menages: l'art de conserver les substances animales et vegetales_ which were to be heated in glass containers and then carefully sealed, he opened a period in history of food entirely novel and in no way anticipated by Willem Beukel's supposed invention of barrelled herring. Canning food is the core of agri-business. Producers of foods to be canned must concentrate on varieties that ship well to cannery and can be canned in recognizable form. Large investment is needed for such operations, and this means ability to find capital for mechanical harvesters to save labor costs. This further focuses production on varieties of food that can be harvested by machines, i.e., those that have been hybridized with the rubber tree. The immense variety of cultivated plants produced over the past ten millennia seems at present to be eroding swiftly. We are depriving ourselves not simply of vast spectrum of flavors and textures, but likewise of an irreplaceable genetic pool which includes potentials for nutrition, for resistance to disease, and much else. In 1975 one of our slogans must be "Don't betray the Neolithic pioneers!"

REFERENCES

1. White L jr: The agricultural revolution of the early Middle Ages. In White L jr: Medieval Technology and Social Change. Oxford, 1962, pp. 39-78.

2. Simoons FJ: Eat Not This Flesh: Food Avoidance in the Old World. Madison, Wisconsin, University of Wisconsin, 1967, pp. 79-86.

3. Jahns M: Ross and Reiter in Leben und Sprach Glaube und Geschichte der Deutschen, 2 vols. Leipzig, 1872, p. 440.

4. Veale EM: The English Fur Trade in the Later Middle Ages. Oxford, 1966, pp. 209-214.

5. Owen C: The domestication of the ferret. In Ucko PJ, Dimbley GW (eds): Domestication and Exploitation of Plants and Animals. London, 1969, pp. 489-493.

6. White L jr: Indic elements in the iconography of Petrach's Trionfo della Morte. Speculum 49:216, 1974.

7. Bradford BB: The application of water-power to industry during the Middle Ages, Ph.D. dissertation, Los Angeles, University of California, 1966. Ann Arbor, Michigan, University Microfilms.

8. Grand R: L'agriculture au Moyen Âge. Paris, 1950, p. 536.

9. Jones JF: The function of food in mediaeval German literature. Speculum 35:80, 1960.

10. Degrijse R, Mus O: De laatmiddeleeuwse haringvisserij. Bijdragen voor des Geschildenis der Nederlanden 21:113, 1966-67.

11. Rosenwein BH: Feudal war and monastic peace: Cluniac liturgy as ritual aggression. Viator 2:129-157, 1971.

12. Bonte M, Von Balen H: Prolonged lactation and family spacing in Rwanda. J Biosoc Sci 1:97-100, 1969.

13. Russell JC: Aspects démographiques des débuts de la féodalité. Annals: Economie, Societies, Civilisations 20:1124, 1965.

14. Brandel F: Civilisation matérielle et capitalisme (XVe - XVIIIe siècle), V 1. Paris, 1967, pp. 139-146.

15. White L jr: The life of the silent majority. In Hoyt RS (ed): Life and Thought in the Early Middle Ages. Minneapolis, 1967, pp. 85-100.

16. Tannahil R: Food in History. New York, 1973, p. 220.

17. Higgs ES, Jarman MR: The origins of agriculture: a reconsideration. Antiquity 43:31-41, 1969.

18. Simoons FJ: Contemporary research themes in the cultural geography of domesticated aniamls. Geog Rev 64:557-576, 1974.

19. Zohary D, Spiegel-Ray P: Beginnings of fruit growing in the Old World. Science 187:319-327, 1975.

20. Watson AM: The Arab agricultural revolution and its diffusion. J Econ Hist 34:9-35, 74-78, 1974.

21. Crosby AW Jr: The Columbian Exchange: Biological and Cultural Consequences of 1492. Westport, Connecticut, 1972.

22. Sauer JD: Changing perception and exploitation of New World plants in Europe. In Chiappelli F (ed): First Images of America. Berkeley, California, 1976 (forthcoming).

23. Langer WL: American foods and Europe's population growth, 1750-1850. J Soc Hist 8:51-66, winter 1974-75.

24. Spencer JE: The rise of maize as a major crop plant in the Philippines. J Hist Geog 1:1-16, 1975.

GEOGRAPHIC PERSPECTIVES ON MAN'S FOOD QUEST

F. J. Simoons, Ph.D.

Department of Geography
University of California, Davis
Davis, California 94616

My "geographical perspectives on man's food quest" are closely linked to Lynn White's views of food and history, for geography, as Herodotus wrote, is the "handmaiden of history." "Geography and history," wrote Ellen Churchill Semple (1) "cannot be held apart without dismembering what is a natural, vital whole." Thus my talk today has a strong historical component. It is also concerned with differences among human cultures as they relate to man's foodways - those customs that regulate man's use of the food resources available to him. In addition my talk considers distributions on the earth's surface, and the spread, through history, of culture traits and complexes that relate to food. Finally my perspective is conditioned by the strong earth science tradition in geography, with its environmental and ecological concerns.

Especially interesting in Lynn White's paper was his consideration of centers of plant and animal domestication, for this topic has been of particular concern to many geographers - most notably Carol O. Sauer and his followers (2, 3, 4). I would, with Lynn White, emphasize the vital roles played by the major centers of domestication - the Near East being the most notable example - in contributing useful plants and animals to the world as a whole. I would also acknowledge the ease with which many plants and animals have spread, especially following European overseas expansion, and the enormous impact particular introduced plants and animals have had on local or national diets. Salaman (5) has chronicled the important place the potato - a South American domesticate - came to assume in the United Kingdom, especially Ireland, William O. Jones (6) and Marvin Miracle (7), in a similar vein, have examined the vital roles played by maize and manioc - also American domesticates - in tropical Africa.

Despite such successes one should not overlook the many other domesticated plants and animals, important as foods in their homelands, which have failed to spread widely or have done so only as curiosities or for nonfood uses. Some of the plants and animals in question derive from the so-called "minor centers of domestication." One of these minor centers is highland Ethiopia, whose best known domesticate is coffee but in which certain food plants were domesticated that have not been utilized elsewhere for food. There is, for example, the tiny-seeded grass *teff* (*Eragrostis teff*), the favorite cereal of the Amhara highlanders who give *teff* better care and soil than other cereals, including wheat (8, 9). Another Ethiopian domesticate is the banana-like plant *enset* (*Ensete edule*) (10), which is cultivated in parts of southern highland Ethiopia and which supports the most dense rural populations in the entire land (11, 12, 13, 14, 9, 15).

Though most scholars, through the publications of Russian botanist Nicolai Vavilov (16, 17) and his students, have long known the importance of the Ethiopian center, recognition of the role of West Africa and the Sudan belt in domestication came much later. In West Africa and the Sudan the oil palm (*Elaeis quineensis*), pearl millet (*Pennisetum* spp.), certain sorghums (*Sorghum* spp.), as well as a group of less-known plants, seem to have been domesticated (18, 19, 20).

What strikes one about Ethiopia and the West Africa/ Sudan zone is how selective has been the diffusion of domesticated plants, how certain plants quite suited to other regions have nevertheless failed to spread. The Ethiopian ensete is a case in point. Even in major centers of domestication, such as the Andean one in South America, though certain domesticates were widely diffused abroad for use as food (the potato, for example), other plants and animals (including llama and guinea pig) were not. Though the Italian Government during the Second World War encouraged the raising of guinea pigs for human food,

Miller Office Supply

in practice set:

do practice A before Friday.

read instruction for A & B

we will work in class together Friday.

there have been few such attempts to use the animal for food outside of its place of domestication (21). Even in the Andes, the acculturation of Indians to the Spanish way had led some to cease keeping and eating guinea pigs. Dr. de Garine's paper will deal with similar situations involving prestige ranking of foods. Such concerns, as well as ones of supply and cost, were involved in the fortunes of horsemeat as human food in France (22).

What the above suggests is that attention must be given not only to the spread of plants and animals, but to the failure of many to spread - and to the quite varied cultural contexts in which domestic and wild plants and, even more so, animals, find themselves. I submit that valuable insights into food use can be gained by viewing human foodways with attention to the individual societies, cultures, and ideological and ecological situations in which they occur. To me, the value of such an approach was brought home forcefully a few years ago when I conducted a seminar on food habits and culture. This was a seminar involving participation not only by geographers, but by sociologists, anthropologists, food scientists, and nutritionists as well. On one evening we arranged a debate on the natural foods movement in which a nutritionist and a food scientist vigorously attacked that movement by identifying nutritional fallacies widely held by natural food advocates. They also noted that natural foods are expensive and often not what they are purported to be. In sheer logic and in documentation of their position, one might assume that the argument was won. But it was not. For the position of the natural foods people went far beyond the facts of nutrition. Theirs was a position that embraced concern with the environment and its destruction, pollution of air and water, desire to escape what were viewed as the conformist structures of our urban-industrial society, and hope of gaining happiness through a simple life involving close, sympathetic ties with other human beings and with nature (but with high-fidelity music, of course).

A century ago, former American Congressman and Ambassador to Italy, George Perkins Marsh, in his book _Man and Nature_ (23) expressed views closely related to those of the natural foods youth to whom I refer. Marsh saw the United States at a critical and in many ways unfortunate juncture in its history, the time following the Civil War when the frontier was being rapidly expanded westward and when its destructive aspects were becoming more and more apparent. Marsh's special love was New England, and he was distressed by the overcutting of forests, erosion of soils, and the resultant uncontrolled flooding of rivers and silting of harbors. American life, Marsh said, was a nomadic one in which men too often simply destroyed resources and moved on. He insisted that we had destroyed our resources almost to the point where our future was threatened. He urged that we remake America into a balanced and well-ordered commonwealth, a land in which man lives in harmony with the natural world. Marsh's views were shunted aside in the unlimited optimism of his day, for the pickings were rich and concern with future generations notably lacking among many pioneers. In time, however, Marsh was returned to the public eye by people like Lewis Mumford (24) who recognized Marsh as the forerunner of the conservation movement in America. Marsh's ideas, it should be noted, are as relevant today as they were a century ago. They are also quite in tune with those of the youth I mentioned above. Those youth, in their views of food and its purity, remain largely unconvinced, and commonly reject the reassurances offered them, even by nutritionist parents. The point is that the food habits of young persons of natural foods persuasion cannot be dismissed by pointing to nutritional fallacies, for their views have a philosophical base and represent a social movement in which concern with the purity and nutritiousness of food is only one manifestation.

Accepting the proposition that food habits should be studied in a sociocultural, historical manner, one may ask for examples of studies done in that spirit. I would note below two contrasting approaches. First there is the so-called "functional approach" used by Audrey Richards (25-56) in studying food among the Bantu peoples of southern Africa. The functional approach in anthropology views each element of society in terms of its function, of the role it plays in the vital ongoing activities of humans. It is the functionalists' view that the social scientist should not focus on historical and cross-cultural parallels, for these remove traits from their proper context within a single culture. Instead, the social scientist should concentrate on the vital relations existing within the individual society and culture. With such an orientation, it is understandable that Audrey Richards' approach, in her classic study of the Bemba of Zambia, deals with food as an interacting part of a sociocultural entity throughout the year. Hers is an approach to food that was not to become prominent in anthropology, but it did lead her to the conclusion that, at least among the Bemba, food occupies a role more important in personal interaction than does sex.

Whereas Richards' functional approach tended to focus on a single society as an object of investigation, what I call the culture historical or geographic approach deals with food habits in broader terms in space and time. Those using the culture historical approach in studies of food habits seek to identify their origins, to trace their spread, to determine their present distribution around the world, and to describe the contrasting positions of such habits in the various societies in which they occur. I include in the culture historical category the fascinating book, Mushrooms, Russia, and History (27), by Wall Street banker R. Gordon Wasson and Valentina, his wife, a pediatrician. In this magnificent two volume book, the Wassons consider mushrooms in terms of their place in European history, especially that of eastern Europe, a region the Wassons note to be one of mushroom lovers, or mycophiles. I would also include in the culture historical approach Xavier de Planhol's book, The World of Islam (28), insofar as it touches on food. Planhol notes that major geographic dislocations occurred in North Africa as a result of conversion to Islam: among them was the Islamic ban on alcoholic beverages which led to a great decline of grape cultivation there, and the Islamic rejection of the pig and pork which led to a decline of swine herding. The latter rejection, moreover, had an impact on vegetation, because afterwards the goat assumed an increased role in the region, for goatflesh was commonly substituted for pork. The goat, moreover, is an aggressive feeder which accelerated destruction of woody vegetation and encouraged erosion, a matter that has attracted attention by historical geographers. Another study in the culture historical category is that of F. S. Bodenheimer, the late Israeli zoologist, on Insects as Human Food (29). Bodenheimer sketches the use of insects as food around the world, and observes that two types of insects - termites and grasshoppers, are most important in human diet generally. He notes that roast termites are richer in protein than roast beef, and urges a more rational view of the custom of eating insects, which is looked down upon by many Westerners, and by educated native peoples.

Turning from the examples above to specific food habits, I would direct attention first to Asia and Africa, continents that have especially serious problems in feeding their populations. In those two regions there are notable differences in the acceptability of particular foods, mainly foods of animal origin. For to eat meat involves the taking of sentient life and with few exceptions entails far stronger feelings than to eat foods of plant origin. There can be problems in the acceptability of meat according to who killed the animal, and of food in general, according to who prepared it and how it was prepared. I would like first to turn to certain of the animal foods that are widely rejected in the area in question, whether by an entire group or by certain individuals within that group.

We are all aware of the unacceptability of pork to the orthodox Moslem and Jew, but it is less well known that in much of Africa - beyond Islamic influence - the pig is also looked upon as an unclean animal, and pork as an unacceptable food. Many of the Africa peoples in question extend their feelings against the pig to related animals, such as the hippopotamus. Around Lake Tana, in Ethiopia, for example, the Wayto are a Moslem people who formerly lived by hippopotamus hunting and fishing. As hippopotamus hunters they were regarded as unclean, and other peoples would not intermarry with them and would not eat freely with them. There is even one account of a Wayto visiting the home of a person of another group. The host, out of courtesy, gave his guest a drink of beer, but after the Wayto left, the host destroyed the vessel.

Another curious food avoidance in Africa is that of chickens and eggs (30). This is quite widespread, not limited just to pastoral groups. Among most of the groups in question, women of child-bearing age are singled out to avoid chickens, eggs, or both. Young girls are commonly permitted to eat such foods before puberty. Old women past child-bearing may eat them, and normally boys and men of all ages may eat chickens and eggs. Some writers have suggested that avoidances such as this, which single out women, were imposed by men in an effort to preserve for themselves the bulk of the animal protein. In any case, today the explanations given by African groups for the chicken and egg avoidance are usually in terms of sex and reproduction. Should a woman eat eggs, it is said, she might become over-sexed and dash off into the forest in search of a man; or she might not be able to bear children; or she might bear children deformed or abnormal in some other way.

Another food avoidance is that of fish, of all fish or all free swimming fish. In Africa, for example, such groups - perhaps best designated "fish avoiders" - are quite numerous (31, 32), though the custom is breaking down rapidly. Some African fish avoiders, it is true, live in desert regions where fish are rare. Others, however, have an abundant fish resource available, yet they refuse to catch and eat fish, and indeed they look down on fish eaters. Fish avoidance is more widespread among the Cushitic peoples in Northeast Africa. The Cushitic groups vary in their attitudes toward fishing and fish-eating. Some reject both practices with what one might call "mild amusement"; other Cushites, however, hold stronger feelings on the matter. Thus, we read of some Cushites who look on fish as unclean water snakes and on fish eating as a disgrace. Most Beja, it is said, refuse fish "even in times of severest want." And the favorite insult used by Somali nomads is "speak not to me with a mouth which eats fish." There is one report that the Walamo of southern Ethiopia in the past even put to death violators of their fish taboo. That, of course, was an extreme case.

Even where fishing and fish-eating occur among Cushitic groups, the practices are looked down upon. Fish among such groups is commonly a low-class food. And fishermen are often viewed with contempt, and referred to in derogatory ways. Thus in Kafa country, fishing is done by the Manjo, a caste of pariah hunters, and the true Kafa do not practice fishing. The Somali refer to coastal fisherman as Rer Manyo, a derogatory term. And the Ababda despise those coastal fish-eating sections of their people "as not having sprung from the tribal ancestor," as not being true Ababda.

It is now known that the abdominal symptoms and diarrhea reported by adults of nonmilking groups when they consume milk, has a biological basis, deficiency of the enzyme lactase in the jejunum. There is, however, no known biological basis for the fish avoidance of the Cushites. Indeed, I have uncovered only a single report of a Cushitic group that believes fish-eating to be detrimental to one's physical well-being in a specific way. This is for the Haruro, who insist that fish is bad for the teeth, that fish-eating would cause their teeth to drop out.

My review of the literature on Africa suggests that African fish avoidance (avoidance of fish in general rather than of particular species) had its origin in the scorn pastoralists commonly display for farmers and their ways. The pastoralists' scorn of fishing and fish-eating presumably was made easy by a sufficiency of protein from their herds. In any case, the pastoralists' fish avoidance could persist as long as their herds were large enough to provide them with an adequate diet. The pastoralists, moreover, were in an excellent position, as people of military prowess and often dominant politically, to lead other groups, including agricultural ones, to take up fish avoidance. If my hypothesis is correct, the prominence of fish avoidance among Cushitic farming groups in Ethiopia is a reflection of a distant pastoral heritage or of pastoral influences, whether in Northeast Africa itself or in an earlier Southwest Asian homeland.

Fish avoidance is also commonplace in India, among certain religious groups (33). Most notable of Indian fish avoiders are adherents of Jainism and of certain Hindu sects, whose rejection is based on the _ahimsa_ concept, nonviolence to sentient beings. The fisherman and fish-eater, according to strict interpretation of that concept, are contributing to the killing of living creatures, an evil act for which they would ultimately be punished. With the spread of Buddhism into the adjacent lands of Tibet, Mongolia, the Far East, and Southeast Asia, ahimsa concerns came to affect human attitudes toward fishing and fish-eating there, too. In Tibet and Mongolia, fish avoidance became commonplace, especially among llamas and devout lay Buddhists. In

Buddhist Southeast Asia, however, avoidance was unusual, and evasion of the concept was the rule, a situation perhaps to be ascribed to the dietary importance of fish, a "nutritional imperative" that overruled commitment to religious belief.

Fish avoidance was also found to occur in Turkestan among settled and pastoral peoples apparently beyond Buddhist influence. These peoples did not reject fish because of concern with taking life, but because they looked on them as dirty creatures. The suggestion was made that ecologic pressures in pastoral nomadism may have contributed to this form of fish avoidance. Once fish came to be avoided by one people, moreover, the custom might be transmitted to neighboring groups or maintained by the first group even though their ecological situation changed. That Homer's heroes in the _Odyssey_ and their linguistic relatives, the early Aryan invaders of India, both may have had negative attitudes toward eating fish, suggests a survival of attitudes gained during a stage of nomadism in the earliest Indo-European homeland, perhaps somewhere in the grasslands from the Ukraine to Turkestan. A third factor that may have contributed to rejection of fish as human food in India was the concept of sacred water and the keeping of fish in sacred bodies of water, whether tanks, ponds, or streams. It cannot, however, be demonstrated that this factor was involved in the development of a general rejection of fish, though it certainly did lead people to avoid catching and eating the fish living in the sacred waters.

BEEF EATING AND THE HINDU SACRED COW CONCEPT

One of the best examples of how powerful the cultural associations of food may become is that of beef and the sacred cow concept of the Hindus. Beef, it should be emphasized, was eaten without apparent prejudice by the ancient Aryans, Indo-Europeans who invaded India about 1500 B.C. and contributed various elements important in modern Hinduism. Beef eating, however, came under strong pressure with the rise of the sacred cow concept, following about 500 B.C. (34).

The practice of eating beef in India did not cease, of course. Certain non-Hindu peoples, including tribal groups and, later, Moslems, continued to eat beef without scruple. In addition, certain low caste Hindu groups persisted in eating beef, but as a result they were not well regarded by other Hindus, and were commonly accused of killing cows in order to obtain their flesh. A ban on cow slaughter, it should be noted, is included in the Indian Constitution of 1950, the only constitution in the world to include such a ban (35). Inclusion of the ban in the constitution, it is true, represented a compromise between the traditional Hindus, who wanted a complete ban on the slaughter of all common cattle, and the secular forces who dominated the Constitutional Convention. The ban, moreover, was included not in the fundamental rights section, but in the "Directives of State Policy," which were guidelines to be implemented by government officials and agencies, but which, in their constitutional form were unenforceable in courts of law. Thus, there was and is no national ban on cow slaughter in India, for the matter was left up to national, state, and local bodies to enact into law. Traditional Hindus in certain local jurisdictions and in certain states have been able to enact into law statutes banning cow slaughter or making it difficult. This, in turn, led to a series of court cases, including three that reached the Supreme Court of India. The Court determined which types of cattle might be killed and under what conditions. Also spelled out were the conditions under which a person might be guilty of cow slaughter, and how severely he might be punished. Eighteen months of rigorous imprisonment was the sentence some judges meted out. In the case of the sacred cow concept, thus, we have a view that links cow slaughter and beef eating with the very roots of Hindu belief, and that finds clear expression in the law.

The sacred cow concept also affects the use of other products of the cow. Whereas it is wrong in traditional Hindu belief to eat the flesh of "mother cow," other products of the living cow play a role in internal and external ritual purification - this in a society where purity/pollution concerns are prominent. The purificatory role of the products of the cow derives from the sanctity of the cow, and is most clearly manifest in the _panchagavya_, the five products of the cow: cow's milk, curd made of cow's milk, clarified butter of cow's milk, cow urine, and cow dung (36). Cow's urine, for example, is widely used to ritually purify places of the house such as the kitchen, that are in special danger of pollution.

The sacred cow concept, it should be noted, is also a matter of ecological concern, at least in the minds of most animal husbandry specialists who have worked in India. That is, the Hindu pressures against cow slaughter, many such specialists hold, lead to a proliferation of cattle, including animals that are old and infirm, that compete with humans for food. One should note that anthropologist Marvin Harris and certain supporters have challenged the view that the sacred cow concept is damaging to India; they approach the problem convinced that techno-environmental factors are overriding in human life and that the sacred cow concept is a useful, "positive-functioned" one that originated because of the vital economic role of the cow in providing fuel, traction, and other services to Hindu Indians. The Supreme Court of India, in its lengthiest of three decisions, reached a contrary position, that the sacred cow concept was, indeed, detrimental to Indian life and well-being. One point made in the Supreme Court decision was that more money was being spent in India per capita to maintain an old cow than to educate a child. Whatever the ultimate resolution of the sacred cow controversy, it should be noted that in order to fully appreciate the food habit of beef rejection, one is drawn into the sacred cow concept, into the vital pollution concerns of Hindus, and into India's legal system and overall ecology.

My focus above is on particular flesh foods which are widely avoided by human groups in Asia and Africa. In addition, one should note that there are concerns in eating that go beyond particular foodstuffs, that relate to more general wishes to avoid ritual pollution, to maintain purity.

Such concerns are especially prominent in the area from northeastern Africa to India. My first awareness of this problem came two decades ago when I was doing field work in northern highland Ethiopia, when on a mule trip for five weeks from the Ethiopian highland to the low Sudan border country. We had with us representatives of three religions: Christian, Jewish, and Moslem. Members of these groups would eat flesh only of animals slaughtered by their own religious group. Thus, the only flesh most of them had on the trip was chicken, for one person could kill and eat an entire chicken, but not an entire goat, sheep, or larger animal. A New Zealand veterinarian had a somewhat similar experience. In his case, he was demonstrating to a group of Ethiopians how to skin a sheep properly. When he was through with his demonstration he offered the flesh to any of the observers who wanted it. None did. It had not been slaughtered by one of their religion.

Purity-pollution concerns reach their most extreme form among Jains and Hindus of India. Concern with impurity is so pronounced and pervasive there that various scholars speak of a "Hindu pollution concept" which is essential for an understanding of how individuals and groups, of whatever sort, are ranked in ritual status. Unlike secular status, which is based on such considerations as ability, education, wealth, and ownership of land, ritual status derives from the relative purity of the individual or group. The pure individual must constantly be on guard, for by contact with an impure person, it is the pure one who becomes polluted and not the impure one who is purified. In such cases, H.N.C. Stevenson (37) has noted, "pollution always overcomes purity."

The forms of pollution have been categorized differently by various scholars, but Stevenson notes that pollution may be permanent or temporary; though temporary pollution can be removed by purificatory ceremonies, if not removed it may become permanent. Much permanent pollution, however, is an inherent condition, as of human groups, parts of the human body, human emissions, plants and animals, and material objects. For human groups, it is differences in permanent pollution that establish their ritual status and their relationships with one another: whether they may intermarry, whether they may eat together, whether they may eat one another's food, and so forth.

Internal pollution, as compared to external pollution, involved penetration by polluting substances, and is by far the more serious of these two and the more difficult to eliminate. Such internal pollution can be developed by eating such food as beef, or eating human excrement, or by eating food prepared by unclean individuals of lower ritual status. This concern with food - with accepting or refusing foods according to who cooks them and how they are prepared - is what Blunt called "the cooking taboo."

Hindus regard raw foods as possessing the greatest purity, and even Brahmins - members of the priestly jatis or "castes" - will accept uncooked grain and other raw foods that have been handled by, or purchased from anyone or almost anyone, including commonly even members of Untouchable castes. Of all raw foods, fruits and vegetables that can be washed and peeled, and nuts protected by a husk or shell, are regarded as the most difficult to pollute; but even they may be refused if their coverings are broken or cut.

In cooking, a raw food is altered ritually, and the Hindu may refuse to eat it if prepared or touched by a person of lower caste, or of different caste. Whatever its derivation, this refusal has led to some interesting customs. One is the preference for Brahmin cooks, for food prepared among them enjoys the broadest acceptability among Hindus. By having a cook of suitable caste, a host can feed a man of high caste even if he himself is of low caste status. Brahmin cooks, notes O'Malley, are also employed in jails to avoid offending caste sensitivities. It is not that the Brahmin cook imparts the purity of his caste, but that he would be expected to pollute it less than would members of lower castes. The cook, on his part, "suffers no ritual hurt" even if the lowliest of people eat food he has prepared.

At religious or quasi-religious affairs Brahmins may be unable to eat the cooked foods of the host because of his low caste, and because he does not have a cook of proper status. In such cases Brahmins may be given raw foods which they cook and eat at the home of the host. A similar custom was observed among a group of Brahmins who in going to a picnic either carried raw food for cooking, or ate only fruits, or took acceptable sweets which they are able to buy at a Brahmin shop in the marketplace.

It is striking, however, that most Hindus do not regard all cooked foods as equally suspect, that in parts of India they distinguish between inferior-cooked or kachcha foods, and superior, or pakka, ones. Kachcha foods, which usually are prepared with water or water and salt but not with ghi or butter, include some of the mainstays of the Hindu diet, such as boiled rice or pulses as well as dry-baked flatbread. Among the ingredients of most pakka foods, by contrast, are milk products, usually ghi, or vegetable oil. Inclusion of the products of the sacred cow make the difference.

There are great variations, from one Indian settlement and region to another, in the patterns of Hindu acceptance of pakka and kachcha foods. The patterns are also undergoing change through time, both because of direct and indirect pressures of urbanization and secularization in India, and because of changes largely independent of these pressures. Overall, however, the pattern is clear: pakka foods, while not as generally acceptable as raw foods, enjoyed broader acceptability than kachcha foods.

Special problems exist for the devout Hindu of status who journeys beyond the confines of his village. In his village, he knows the castes and the acceptability of their foods for him. He can also readily prepare ritually pure foods at home. In traveling, however, he must take special precautions to maintain his ritual purity. He may take along foods prepared at home, or he may buy raw foodstuffs that he himself or some member of his family can prepare en route. He may, if he buys foods along the way, limit his consumption to fresh fruits or other raw foods that he can wash and perhaps peel. Or he may buy pakka foods from a vendor and consume them. He would, however, be much less likely to eat kachcha foods. Because such concerns are common, a variety of raw foods are usually available for sale to travelers in India, and a wide range of pakka foods as well. There are, nevertheless, numerous violations of the ideal pattern of food avoidance when people are traveling. Certain Indian observers note that the old food restrictions, including those pertaining to consumption of pakka and kachcha food, no longer apply generally "to the conditions of city life and travel." There now exist two standards of behavior: that which people follow in their villages and that which they follow when in cities or away from home. It should be noted, however, that less than 20 percent of India's people live in urban areas.

The above foodways seem to have roots solely in human culture. In the case of milk, however, the group differences that exist involve not only cultural attitudes but a genetic component as well, in ability to digest lactose, milk sugar.

Dairying seems first to have been practiced about 4000 B.C. in the Near East and North Africa (38). The habits of milking animals and using their milk then gradually spread across Europe, Asia, and Africa, but even today they have not been taken up by all peoples in those regions. As of 1500 A.D., the date after which European overseas expansion came to modify so greatly the traditional ways of life in Asia, Africa, and the Americas, dairying was completely absent among American Indians and Eskimos. Though most of these, of course, had no animals suitable for milking, the Andean peoples of South America had llamas and alpacas that conceivably could have been milked, but were not. In Africa, dairying and milk use were absent in about one-third of the continent, including the shores of the Guinea Coast of West Africa, the Congo Basin, and an area of East Africa extending across Mozambique to the Indian Ocean (39). Elsewhere in Africa, milking was practiced universally by people who had suitable domesticated animals. Such peoples consumed varying amounts of milk - in some cases, as with pastoralists such as the Fulani of Nigeria, consumption was high and in other cases, as among various agricultural peoples, it was small.

In any case, there are clear indications that dairying and milk use had spread southward from Kenya into southern Africa quite late (following the beginning of the Christian era) and that the peoples of those regions, now mainly Bantu in speech, only recently took up the habits. There has been much speculation about the failure of dairying and milk use to spread throughout Africa. Some have pointed to the cultural impediments to dairying: to the views expressed by native nonmilking groups that milk is a dirty bodily secretion coming from an animal, a white substance unsuited for human consumption, and that the practice of milking is an unnatural one. The African nonmilker also insisted that if he consumed milk he would get stomach gas, cramps, diarrhea, and even vomiting. Altogether the above came to be described as "the nonmilking attitude," a complex of views advanced by the nonmilker to explain his refusal to practice dairying and to use milk. Others dealing with milking in Africa focused on environmental impediments to the spread of dairying, especially those which excluded the vigorous pastoral groups around the fringes of the nonmilking zone who might have been expected to introduce the milking habit there. The principal environmental impediment was tsetse-borne sleeping sickness, that disease which affects common cattle quite severely and largely excludes them from the nonmilking zone. The problem of sleeping sickness is one that has attracted much attention because of its detrimental effect on the life of humans and their domestic animals. There is a vast and growing literature on sleeping sickness, and it is clearly one that must be taken into account in explaining why the African zone of nonmilking was able to persist into so late a historical period.

In Asia, where the nonmilking zone embraced all of Southeast Asia from Burma eastwards, as well as most of the Far East, including China, Korea, and Japan, native peoples gave similar reasons to those of African nonmilkers for not practicing dairying and using milk (40). Like his African counterpart, the nonmilker of Eastern Asia also claimed to get sick on consuming milk. And he had various additional explanations, deriving from the ahimsa or nonviolence concept, and from other cultural views. In explaining the persistence of the zone of nonmilking in eastern Asia, where such dairying peoples as Mongols and Tibetans were close by, some writers have spoken of environmental impediments - including the low nutritional value of grasses in the S. E. Asian tropics - and others have suggested that dairying simply doesn't fit well into the intensive agriculture that characterizes so much of the Far East. These explanations gloss over the fact that whatever the impediments of environment and ecology, Indian migrants in Southeast Asia were able to identify and cleverly exploit ecological niches suitable for dairying, and that their success is based largely on the demand created by other Indian settlers in the region.

Indians in Southeast Asia today still are the primary market for Indian dairymen.

The answer to the failure of dairying to spread throughout Asia and Africa does not involve a single cause, but a series of factors from culture, environment, and ecology. In addition, medical research starting a decade or so ago has identified another factor that is involved: that most persons after about age three in the nonmilking zones have low levels of intestinal lactase, the enzyme that hydrolyzes lactose into glucose and galactose which can readily be absorbed by the body. The individual with low lactase can hydrolyze only small amounts of lactose. The unhydrolyzed lactose, moreover commonly leads to stomach gas, distension, cramps, and diarrhea in the malabsorber (41). Thus, the symptoms described by the African and Asian nonmilkers, advanced by them as explanations for their refusal to consume milk, may well have had a biological basis and were not, as many of us suspected, merely psychosomatic in origin.

Recent research has confirmed an interesting geographic pattern of high and low incidences of malabsorption (42). Though my time remaining is too short to go far into the matter, I would like to note that all unmixed groups from within the traditional zones of nonmilking have high incidences of primary adult lactose malabsorption, that is from 70 to 100% of the individuals studied have proved to be malabsorbers. This, it should be emphasized, is the same condition that prevails in other land mammals, and we believe it to be the normal pattern found among early men. How human groups within the milking zone came to have low incidences of malabsorption - and only certain of them have such low incidences - remains a matter of controversy. I suspect, however, that under certain conditions where milk was important in diets that otherwise were inadequate nutritionally and where milk was not processed into low lactose forms - the aberrant individual with high levels of intestinal lactase persisting after weaning would be favored in the struggle for survival. Thus, in a classical Mendelian way the trait of lactose absorption would come to predominate in the groups in question. If this so-called culture historical hypothesis (43) is correct, we are dealing with a food habit - the consumption of milk - which under certain conditions led to significant genetic

REFERENCES

1. Semple EC: Influences of Geographic Environment. New York, Holt, Rinehart and Winston, 1911.

2. Harris DR: New light on plant domestication and the origins of agriculture: a review. Geogr Rev 57:90-107, 1967.

3. Isaac E: Geography of Domestication. Englewood Cliffs, New Jersey, 1970.

4. Sauer CO: Agricultural origins and dispersals, Bowman Memorial Lectures, 2. New York, American Geographical Society, 1952.

5. Salaman RN: The History and Social Influence of the Potato. Cambridge, Cambridge University Press, 1949.

6. Jones WO: Manioc in Africa. Stanford, California, Stanford University Press, 1959.

7. Miracle MP: Maize in Tropical Africa. Madison, Wisconsin, University of Wisconsin Press, 1966.

8. Ciferri R, Baldrati I: Il 'Teff' (Eragrostis Teff). Cereale da panificazione dell'Africa orientale Italiana montana. Vol. 2 of I cereale dell'Africa Italiana. Florence, Regio Instituto Agronomico per l'Africa Italiana, 1939.

9. Simoons FJ: Northwest Ethiopia: Peoples and Economy. Madison, Wisconsin, University of Wisconsin Press, 1960.

10. Cheesman EE: Classification of bananas: I. The genus _Ensete_ Horan. Kew Bull 2:97-106, 1947.

11. Bezuneh T: The role of musacae in Ethiopian agriculture. I: The genus _Ensete_. Acta Hort 21: 97-100, 1971.

12. Copertini S: Il 'pane' di fecola di banano abissino. L'agricoltura coloniale 32:444-446, 1938.

13. Shack WA: The Gurage: a people of the Ensete culture. London, Oxford University Press, 1966.

14. Simmonds NW: Ensete cultivation in the southern highland of Ethiopia: a review. Trop Agri 35:302-307, 1958.

15. Smeds H: Theensete planting culture of Eastern Sidama, Ethiopia. Acta Geogr 13:1-39, 1955.

16. Vavilov NI: Studies on the origin of cultivated plants. Leningrad, Institute of Applied Botany and Plant Breeding, 1926.

17. ___________: The origin, variation, immunity and breeding of cultivated plants. Translated from the Russian by Chester KS. Chronica Botanica 13: Nos. 1-6, 1949-1950.

18. Harlan JR: Agricultural origins: centers and non centers, Science 174:468-474, 1971.

19. Murdock GP: Africa: Its Peoples and their Culture History. New York, McGraw Hill, 1959.

20. Portères R: Berceaux agricoles primaires sur le continent Africain. J Afr Hist 3:195-210, 1962. (Translated as Primary Cradles of Agriculture in the African Continent, Fage JA, Oliver RA (eds) Papers in African Prehistory. Cambridge, Cambridge University Press, 1970.

21. Gade DW: The guinea pig in Andean folk culture. Geogr Rev 57:213-224, 1967.

22. ___________: Horsemeat as human food in France. Ecology of Food and Nutrition, in press, 1975.

23. Marsh GP: Man and Nature; or Physical Geography as Modified by Human Action. New York, Charles Scribner, 1864.

24. Mumford L: The Brown Decades: A Study of the Arts in America, 1865-1895. New York, Harcourt, Brace, 1931.

25. Richards AI: Land, Labor and Diet in Northern Rhodesia. London, International Institute of African Languages and Cultures, 1939.

26. ___________: Hunger and Work in a Savage Tribe. Blencoe, Illinois, The Free Press, 1948.

27. Wasson, VP, Wasson RG: Mushrooms, Russia, and History. 2 vols. New York, Pantheon Books, 1957.

28. dePlanhol S: The World of Islam. Ithaca, New York, Cornell University Press, 1959.

29. Bodenheimer FS: Insects as Human Food. The Hague, W. Junk, 1951.

30. Simoons, FJ: Eat Not This Flesh. Madison, Wisconsin, University of Wisconsin Press, 1961.

31. Lagercrantz S: Forbidden fish. Orientalia Suecana 2:3-8, 1953.

32. Simoons, FJ: Rejection of fish as human food in Africa: a problem in history and ecology. Ecol Food Nutr 3:158-164, 1974.

33. ___________: Fish as forbidden food: the case of India. Ecol Food Nutr 3:185-201, 1974.

34. Brown WN: The sanctity of the cow in Hinduism. Madras Univ J 28:29-49, 1957. (Reprinted in Econ Wkly, Calcutta 16:245-255, 1964.

35. Simoons, FJ: The sacred cow and the constitution of India. Ecol Food Nutr 2:281-296, 1973.

36. ___________: The purifactory role of 'the five products of the cow' in Hinduism. Ecol Food Nutr 3:21-34, 1974.

37. Stevenson HNC: Status evaluation in the Hindu caste system. Jl R Anthrop Inst 84:46-65, 1954.

38. Simoons, FJ: The antiquity of dairying in Asia and Africa. Geogr Rev 61:431-439, 1971.

39. ___________: The non-milking area of Africa. Anthropos 49:58-66, 1954.

40. ___________: The traditional limits of milking and milk use in southern Asia. Anthropos 65:547-593, 1970.

41. Kretchmer N: Lactose and lactase. Sci Am 227:70-78, 1972.

42. Johnson JD, Kretchmer N, Simoons FJ: Lactose malabsorption: its biology and history. Adv Pediatr 21:197-237, 1974.

43. Simoons FJ: Primary adult lactose intolerance and culture history. Mod Probl Paediatr 15:125-141, 1974.

44. Borgstrom G: The Food and People Dilemma. Belmont, California, Duxbury Press, 1973.

45. Portères R: Vieilles agricultures africaines avant le XVIeme siècle. Berceauz d'agriculture et centres de variation. L'Agronomie tropicale 5:489-507, 1950.

46. ___________: Géographie alimentaire-Berceaux agricoles et migrations des plantes cultivées en Afrique intertropicale. Paris, C.R. Soc Biogéographie 239:16-21, 1951.

AVAILABILITY OF FOOD

E. M. Ojala, D. of Philosophy

Food and Agriculture Organization
of the United Nations
Via delle Terme e Carcalla
Rome, Italy

I am pleased to be with you today and to contribute to this multidisciplinary study of food, man, and society. This theme reflects the duality of approach that is essential if an improvement in the nutritional wellbeing of humanity is to be achieved. It is necessary to examine the food-man equation not only in terms of aggregate supplies for a total number but also in terms of constraints on the availability of food to the individual-constraints which may be imposed by society. I shall attempt to pay particular attention to this second, sometimes overlooked, aspect, for it is man's society and his place in it that, to a large extent, determine whether he has access to sufficient food, or is to be counted, along with his children, among the hordes of the hungry.

I note that my contribution on the availability of food is placed early in the deliberations of this Congress. This may well be appropriate in symbolizing the key role that those whose task it is to make food available must play in the better nourishment of mankind. I propose, therefore, to present to you some of the salient facts concerning both the struggle to produce enough food and the distribution puzzle of ensuring that the availability of food matches better the dispersion of the human need. I hope that this will serve as a worthwhile contribution to the broad approach reflected in the theme of the Congress.

FOOD PRODUCTION AND POPULATION

The logical starting point in a review of food availability is to consider whether food is being produced in sufficient quantities and whether it can be produced in the future in sufficient quantities to meet the fast-rising number of mouths to be fed. The Food and Agriculture Organization of the United Nations (FAO) is constantly engaged in reviewing available information to throw light on these questions.

Views about the ability of the world to produce sufficient food for its inhabitants tend to be influenced by performance in the most immediate past. Opinion sways from optimism to pessimism according to the size of recent harvests. It is wise to take a longer perspective in food production and population growth. For this discussion, let us consider the period from 1952 to 1974.

The human population seems to have grown in the last twenty years at a rate unprecedented in history. This is due to a major acceleration in the population growth rate of the developing regions resulting from rapidly declining mortality rates and static or even rising fertility rates, both of these due to a welcome improvement in health facilities and education in these regions. These developments have more than counterbalanced the continued slowing down of population growth in the industrialized countries. At this time the annual addition to the population is around 80 million. The population that had to be fed in the middle of the period reviewed, 1962-1970, is only half of that which may be expected to need feeding by the end of this century.

In view of this demographic situation it is no mean achievement that world food production expanded faster than population in both the 1950's and the 1960's. Per capita food production, therefore, was improving during this period. However, a warning note is struck by the fact that the improvement in per capita production was slowing down in the 1960's compared with the earlier period. In the developing market economies the improvement in per capita food production in the 1960's was only 0.5 percent a year, compared to 0.7 percent a year for the previous decade. This slowing down occurred despite major efforts to increase food production in many countries.

Although the growth rates of food production in developing regions compare favorably with those in the developed countries, the growth of per capita food production was much higher in the latter due to the much lower, and declining, rates of growth in the population. In the 1960's, indeed, food production was held back in certain developed countries in order to avoid the accumulation of excessive surpluses.

The period 1970-74 covers the beginning of the Second United Nations Development Decade. It has had, as far as food production is concerned, a most inauspicious start. Bad weather on a widespread scale has affected food production, especially in 1972 and in 1974. In 1972 the absolute level of food production fell in the developing countries and in the world as a whole. This decline at the world level was probably the first since the end of the Second World War. The fall in food production affected most of Europe, Russia, and North America and was particularly severe in Asia and the Far East. In 1973 it was the turn of Africa and the Near East to be hit - in this case by droughts, although production in other regions in general recovered from the slump of 1972. Special efforts were made in most countries to produce a bumper food crop in 1974 but again the weather intervened, abetted now by the shortage and high price of fertilizers. Once again food production decined in both North America and the Far East.*

As a result of these difficulties, per capita production of food has declined in the 1970's in the developing market economies by more than one percent per year. In Africa the decline in food production has been on the order of two percent per year. Asia and Latin America also showed declines of more than one percent per year. As a result food production per capita at the world level has been almost static since 1970. The race between food

*Data on food production - FAO; population - U. N.

production and population growth rates is no longer being won and efforts to remedy this situation must be intensified.

The seriousness of the total supply situation, particularly in relation to the poor harvests in the 1970's, is reflected in the situation with regard to cereals. In the 1960's there were large surplus stocks of cereals, especially in North America. Indeed, these surpluses were an embarrassment to those holding them, concerned as they were about the effect which such stocks, or even their disposal for food aid, might have on prices in the commercial markets. Now, however, cereal stocks are minimal and insufficient to guarantee any reasonable level of world food security. We are currently in the position that the very existence of cereal stocks at all is dependent on the size of the current year's crop. The Director General of FAO has repeatedly drawn attention to the dangers in the present situation, and recently put forward proposals for a system of world food security towards which the nations are gradually and cautiously feeling their way. The system is intended to ensure that a certain minimum amount of food is available from year to year, through a coordinated network of national stocks.

FOOD AVAILABILITY AND PHYSIOLOGICAL REQUIREMENTS

So far I have been discussing the availability of food in terms of changes in the rate of production compared with population growth, for the world as a whole and for major regions. This approach is useful but too general for our purpose.

I turn now to look at the matter in terms of the amount of food available for consumption at the national level. Domestic production is the main determinant of supply, but the balance between imports and exports of foods is also taken into account. FAO builds up this information, food product by food product, separately for every country in the world. The result for each country is what we call a national food balance sheet. It shows the quantity of each food available for consumption by the inhabitants of the country in one year, on the assumption that the supply is evenly distributed. Using food composition tables, we convert the quantities of food consumed per head per day into nutrient elements such as calories, proteins, and fats. This enables a comparison to be made at a national level between average intake of dietary energy and the calculated physiological requirements based on FAO/WHO (World Health Organization) standards. A good deal of estimation is entailed, but these data are the most comprehensive set available.

In the developing regions the total supply of food is five percent below the level of physiological requirements, with major deficits in Africa and Asia. This deficit must be much greater today after several years of declining per capita production in those areas.

In the developed world the dietary energy supply available in calories per capita was 43 percent higher than in the developing world, and more than 20 percent above the level of physiological requirements.

In two developing regions - Latin America and the Near East - the dietary energy available in 1970 slightly exceeded the physiological requirements. Excluding these, it may be said that in 1970 over 2,000 million people, or some 60 percent of humanity, lived in regions where the overall food supply was insufficient.*

Before considering the differences among developing countries and the distribution of food within countries, it is necessary to take account of the nature of the national diet and the national food production system as a factor in the availability of food.

PATTERNS OF FOOD AVAILABILITY

There is a great diversity among countries not only in levels of food consumption, but also in patterns,

---

*Source: FAO, UN, WHO.

reflecting the natural resources and the nature of the national food production system, especially as regards the role of livestock.

In one group are the great meat eaters, New Zealand and the United States, consuming over 110 kgs. of meat a year per capita, along with 90-100 kgs. of cereals, 50-70 kgs. of potatoes and some 50 kgs. of sugar. Argentina, although a poorer country, is in this group of meat consumers, being like New Zealand a relatively low-cost producer of meat on the range; but it uses less sugar per capita and makes up calories with starch roots and cereals. France is not far behind this group in meat consumption. Other developed countries such as Italy, Greece, and Spain have about half as much meat and sugar available per head, and make up with cereals or, in the case of Spain, with starchy roots. The populations of Venezuela, Brazil, and Mexico, at lower income levels, consume much less meat - 20-40 kgs. per head, but have quite a lot of sugar. More typical developing countries, such as the Philippines and Morocco, can consume only one fifth or one sixth as much meat as the developed countries and much less sugar, but they consume almost twice as much cereals. Similar to these is India, except that average meat consumption is only 1.5 kg. a year. Typical of low income, humid African countries is Ghana, with a high availability of starchy roots and relatively small amounts of meat and sugar. In Japan, the average diet for resource reasons is more akin to developing countries, being low for meat and high for cereals. Despite the high availability of cereals or root crops in developing countries, the total food supply is usually insufficient. It is only in these circumstances, where average dietary energy supply is deficient, that protein supply may be deficient also. The main exceptions are found in the limited areas where the staple food is very low in protein - for example, cassava - and in the case of children who may be unable to eat enough of a bulky, staple food to satisfy their protein requirements. Otherwise, adequate availability of calories usually implies adequacy of protein supply.

The contrast between food energy availability per head in high income and low income regions is made more striking if the calories used in the production of livestock products consumed as food are added to those obtained directly from vegetal sources. On this basis, in terms of original calories, the availability of food energy in the great meat-eating zones exceeds 11,000 calories per head per day, as against less than 3,000 in many countries of Asia and Africa. This calculation illustrates the built-in reserve of energy availability in the diets of the developed countries which is not present in most developing countries. This reserve is most significant politically in the cases where the livestock system is based predominantly on the feeding of cereals rather than the grazing of rangelands. In Europe, for instance, the gross availability of cereals is 400-500 kgs. per capita per year, of which 300-400 kgs. are fed to livestock. In some countries, such as Denmark, Canada, and the United States, two to three times this availability of cereals per head is used for producing more expensive foods from livestock. In typical developing countries, on the other hand, it is a struggle to attain an average total availability of cereals of 200-250 kgs. per capita, most of which has to be consumed directly as the main component of the diet.

It must be evident from this analysis that the accident of an individual's birthplace is a major factor in determining the availability of food to him. A person born into an industrialized society, where agriculture has become so productive that a large proportion of the agricultural land can be used to grow feed for livestock, can look forward to enjoying an abundance or super-abundance of food and a highly diversified and palatable diet, in which livestock products are a major element. Twice as many human beings are born into low-income, agricultural societies in the developing regions. Here, 60 to 90 percent of the population are farmers, but agricultural productivity is low and the availability of food is limited almost entirely to direct home-grown products of the land, mainly cereals or starchy roots, and there is not enough even of these monotonous foods to go around, let alone to store against a bad season.

DISTRIBUTION OF THE AVAILABLE FOOD WITH A COUNTRY

It is indisputable that even in the most efficient society, under whatever political creed, perfect distribution of food according to need is an ideal that has not yet been attained. In most countries the inequalities in food distribution are glaring and degrading in their consequences. Shortage conditions have to be extreme enough to threaten national survival before a society resorts to full scale rationing, although there are a few developing countries which have introduced an official distribution system for certain basic foods parallel to the market channel. In the absence of rationing, total food availability in a country has to be significantly in excess of aggregate individual requirements, in order to allow for the inevitable over-consumption and waste by the more fortunate members of the society.

There is no set standard as to what excess margin is desirable or essential, but it has been observed that in the developed countries the availability of dietary energy is consistently some 20 percent above the calculated requirement of the population; and it may be assumed that in these countries practically the whole population has enough to eat. Some experts consider that a maldistribution margin of 10 to 15 percent at the national level is adequate but necessary, before the least well fed sectors of the population can be assumed to have enough.

There are very few countries in the developing world in which the over-all supply of food is sufficient to allow for maldistribution. If we take a conservative standard of availability of food equal to 110 percent of calculated requirements, there were only 14 developing countries out of 96 who attained this level in 1970. Of these fourteen, seven were in Latin America, three in Africa, and four in Asia.

At the other, more unfortunate end of the scale there were no less than 23 developing countries in 1970, with an availability of food at the national level that amounted to less than 90 percent of their physiological requirements. Of these twenty-three, thirteen were in Africa, six in Asia, and four in Latin America.

In all countries below the 110-115 percent level, there are bound to be sections of the population that are short of food. In countries below the 90 percent level, the amount of chronic hunger and malnutrition is beyond the imagination of people in developed countries.

Unfortunately, there are few data from which a quantitative picture of the distribution of food within a country can be derived. Indeed the lack of adequate data bedevils the detailed consideration of food availability in all its distributional aspects. To some extent we are still groping in the dark when we attempt to answer the questions: "Who are those who are short of food, how many are they, and where are they located"?

For most people who are short of food the reason is that they do not have the means to avail themselves of it. For those whose food supply is mainly dependent on their own farming efforts - and these still constitute almost half of humanity - the availability of food is determined by the amount of land to which they have access, and their security and productivity in the cultivation of it. For the urban dwellers and the landless rural population, food availability does not necessarily depend on adequate supplies of food in the markets, but rather on whether these people have the means to purchase the food. For people without jobs and without money, food is unavailable even in times of plenty.

Another group for whom the problem of food availability may be particularly acute is the younger section of the community. This is largely a matter of the distribution of food within the family. If food is short or the means to acquire it inadequate, the working adults must meet their minimum needs in order to continue working or searching for work. They may draw, unwittingly, from the share of the family's food needed by the child, for a retarded growth rate and poor development are not immediately obvious in the child, and in any case are less critical to the family's survival. Data from a recently conducted survey in the Philippines show that children below 12 years of age in all the classes of the population studies in two provinces have average energy intakes well below the recommended levels.

Not only may the child do poorly in the allocation of the food available, but his special needs for certain nutrients may not be met by the diet offered. A hastily-weaned child, and even an older child, may starve in the midst of plenty - plenty at least in terms of the availability of plantains and root crops, whose protein content is so low that he cannot ingest enough to meet his protein needs.

The danger that maldistribution within the family will seriously affect the children is, of course, greatest in the poorest families. Hence, the main general factor determining the availability of food to individuals within any given society is income.

The link between income and food availability can be clearly seen in data recently made available to FAO by the National Sample Survey of India. The data for rural and urban population in the states of Orissa and Bihar relating to 1971-72 show that the dietary energy available to households according to their total expenditure may be taken as a proxy for their relative position with regard to income or wealth (1).

I am not concerned, for the moment, with the possible errors in the absolute figures, but rather the trends across income groups. The pattern for these and other Indian states and for rural and urban areas is the same; food availability increases with wealth. The food available to the poorest sections of the community is less than one-half of that available, not to the 'rich' groups but to medium-income groups who seem to meet their food needs.

It can also be seen from these tables that the rural poor fare slightly better than the urban poor. The figures for Orissa are typical: the two poorest rural groups had energy availability of 1,320 and 2,000 kcals per capita per day, whereas the equivalent urban groups had 1,150 and 1,670, respectively. The figures show quite clearly that even if food is available in reasonable quantities for those with limited but adequate resources, it is certainly not available to the very poor. Similar data are available for a small number of other developing countries showing the same kind of relationships. Studies have also been made in developed countries.

Although such data are sparse, they do provide a basis - though an inadequate one - for attempting to distribute the population of the world by levels of food availability, and thus to assess the number of malnourished people in the world.

Using a new methodology, FAO recently made such an estimate. The method aims to assess, by an analysis of probabilities and on rigorous and conservative criteria, the numbers of people whose food intake is below the minimum required to provide energy for body maintenance.

The estimate comes to the appalling figure of 400 to 500 million people suffering from severe protein-calorie malnutrition, and that excludes China for which there is no comparable basis for this assessment. The figure relates to 1970 and is subject to marginal doubts and errors, but it points to a human condition which should not be tolerated in this century - 15 percent of mankind without enough to eat. Some thirty million of these people are in developed countries, but the rest eke out their lives in the developing regions of the world. The problem is particularly acute in Asia and Africa, where 30 percent and 25 percent of the respective populations are affected.

It must be stressed that these hungry millions constitute a "normal" state of affairs in the present-day world. As already indicated, they are the poorest section of the world's population. Most of them are rural people - nomads, small farmers, and landless workers and their families, whose poverty and hunger result from their low productivity in food production. Others are urban dwellers, without adequate or secure employment. A large proportion of them are young children.

If, by definition, their food intake is below the minimum required to provide energy for body maintenance, how do these people survive? By reducing work and other activity, or by losing weight, or by not growing or learning if they are children - in other words, by adjusting to a sub-standard existence, and becoming increasingly vulnerable to disease and death.

In seasons of crop failure, or periods of food shortage, high prices, economic distress, recession, unemployment, the food situation of these groups deteriorates sharply, with the threat of starvation, and hundreds of millions more who normally manage to attain the level of nutritional adequacy are added to their number. This has, in fact, happened in the last few years.

To summarize crudely the stage reached in the availability of food to humanity, it may be stated that one-third of mankind - mainly people who live in the developed countries where agriculture has been made very productive - have more than enough to eat, and, in fact, enjoy a diversified diet that is even extravagant in the consumption of food energy. Fifteen percent of mankind are regularly hungry in the sense described above, and are never far from the threat of starvation. Most of the remainder live on rather monotonous cereal or root-based diets - a considerable but unknown proportion of them so little above the subsistence margin that they are forced to adjust to calorie deficits in periods of stress, through illness, disease, or death.

FUTURE PROSPECTS FOR FOOD AVAILABILITY

What then are the hopes for the future? Can food, and the right food, be made available to the hundreds of millions now in need and the millions yet to be born? Can we guarantee that future generations of children will not be condemned to inadequate nutrition, infection, and mental as well as physical stunting, if not death?

Extrapolation of production trends and projections of demand made by FAO show that in the developing market economy countries the demand for food is likely to expand at 3.6 percent per year while food production is rising at only 2.6 percent. The gap is widest in Africa with demand estimated to grow at 3.8 percent but food production at only 2.5 percent. Unless steps are taken to alter these trends there would be by 1985 a very substantial imbalance between production and demand in all developing regions. It has been estimated that by 1985 the developing market economy countries would have a net deficit of cereals of 85 million tons per year and, if the cereal-exporting countries such as Argentina and Thailand are excluded, the projected deficit rises to 100 million tons a year.

Moreover, these comparisons of production and demand do not take account of the need to feed the hungry who because of their poverty are unable to create a demand in the economic sense. Even if production matched the demand estimates, the inadequate availability of food for those most in need would persist.

The developed countries have the food production capacity to make up the projected deficits of the developing regions, provided the problems of logistics and financing can be solved. But the true solution lies in the dramatic acceleration of food production in the developing countries themselves.

Ever since Malthus the question of whether food can be produced in the necessary quantities has been debated. Man in his ingenuity seems capable of solving the technological problems associated with increasing production at the required rate as long as he has the will to do it, but a higher growth rate in food production alone will not bring about adequate food availability for all. What is needed for this is a change in the fundamental relationship between man and his society.

All evidence suggests that there remains an enormous potential for producing enough food to feed the much larger population that, as projected with some reliability, will be in existence by the end of this century. During that time it may be hoped that natural forces coupled with government policies will bring about a reduction in the population growth rates that, if unchecked, will intensify the problem of food availability; but production policies will have to be accompanied by consumption and distribution policies.

A selective approach to agricultural development in which the resources are concentrated in the areas and on the farmers best able to utilize them will not achieve the desired result. Development is failing to reach the small farmer and the landless laborer. These are the hungry ones. In countries with low population densities and adequate land resources a redistribution of land, as politically feasible, and the settlement of new lands will be necessary and vital factors in remedying the situation. In densely populated countries the problem is more difficult. Redistribution, *per se*, will merely create an even larger number of nonviable small holdings. It is necessary to make as full use of the human potential as of the natural resource potential, and the achievement of the former may require more radical changes than are necessary for the latter. Bold innovations in the structure of agricultural production will be needed to facilitate and accelerate the adoption of known improved technologies and modern inputs in food production, without increasing rural (or urban) unemployment. Such innovations may take many forms, depending on national preferences, but an essential feature must be an integrated approach to rural development, including simultaneous investments in agricultural and nonagricultural production.

For countries that cannot, or should not, seek self-sufficiency in food, the availability of food will be linked to their export earnings and the climate of international trade. Major reforms in the world trading systems will be necessary to bring about a more equitable balance of trade for the poor countries.

In summary, it must be realized that at the root of the problem of food availability is the inequality in income, wealth, and economic power both at the national and international levels. The problem will not be solved until the determination is there, both in the governments of the world rich and poor, and among the planners and administrators within each country. The availability of food for humanity will not, in the last analysis, depend on the efforts of the agriculturalist, but on the political and social will of the governors of the peoples of the world.

REFERENCE

1. The National Sample Survey Twenty-Sixth Round: July 1971-June 1972; Ministry of Planning, Government of India, April 1975.

CLINICAL MANIFESTATIONS OF MALNUTRITION

E. M. DeMaeyer, M. D.

World Health Organization
1211 Geneva 27
Switzerland

The symptomatology of specific deficiency diseases and of protein-energy malnutrition is too complex to be used as such in the assessment of their importance from a public health point of view. Only those signs which can easily be identified or accurately measured and are observed with a high degree of constancy can be used for such a purpose. Besides those direct indicators, a crude assessment of the situation can also be obtained by the analysis of vital statistics, food consumption surveys, and the ecological background. As an example, we shall review in the present paper the various criteria that can be used in the assessment of protein-calorie malnutrition (PCM) and of avitaminosis A, two of the most important deficiency diseases existing at present in the world.

PROTEIN-CALORIE MALNUTRITION

Clinically, children may be classified as <u>kwashiorkor</u> (edematous form), <u>marasmus</u> (dry form) or <u>marasmic kwashiorkor</u> (intermediate form with edema). The symptomatology is rather complex and may include varying degrees of growth failure (weight and height), gastro-intestinal disorders (loss of appetite, vomiting, diarrhea), lesions of skin and hair, mental changes, and edema in the case of kwashiorkor and of marasmic kwashiorkor. There are also various signs of avitaminosis and changes in the blood, urine, and cell composition; hypoalbuminemia is one of the earliest and most characteristic changes in kwashiorkor and marasmic kwashiorkor. It is obvious that this symptomatology is too complex to be used as such and that simple criteria represented by constant and easily observable signs must be identified for assessing the importance of malnutrition. The following criteria can be considered for such a purpose.

Clinical Signs

Edema. The presence of bilateral edema is an almost unmistakable sign of moderate to severe protein-energy malnutrition in children; the causes of erroneous diagnosis are few. In pregnant women, the chances of errors are greater but do not affect significantly the results if edema is detected in a large number of women. It is a sign of advanced protein-energy malnutrition and as such will not be observed in a significant percentage of the vulnerable population (preschool age children, pregnant and lactating women) unless the nutritional situation is gravely affected or a fairly large sample of population has been examined. Attempts have been made to standardize the test for detection of edema, although this has never been fully achieved. The presence of edema is characteristic of the kwashiorkor or marasmic kwashiorkor forms of PCM but fails to identify the marasmic forms. There is, therefore, a tendency to underestimate the importance of the problem. It is, nevertheless, a useful indicator, especially in emergency situations when no other one is available; it has been used with success during the drought in the Sahel.

Anthropometric Measurements

Weight for age. This is the simplest and most common measurement for the assessment of growth. A scale is required which must be checked and adjusted with known weights; ideally, this should be done every day before any measurements are made. For the assessment of the nutritional status, the actual weight should be compared with a reference weight, i.e., the weight of a "normal child" of the same age. There have been diverging opinions as to which reference growth figures should be used, i.e., national or international ones. There is much to say in favor of an international reference growth curve since, in most cases, the influence of environmental and especially of nutritional factors is greater than that of genetic factors in determining the growth of children. There are few national or local growth curves available, and the time and money required to determine one are often out of proportion with the benefits.

It is on the basis of this comparison that the classification of Gomez for malnutrition has been established. Three degrees of malnutrition have been defined, i.e., first degree for children with a weight between 75 and 90% of the reference child of the same age; second degree for children between 60 and 75%, and third degree for children below 60% of the reference weight. In this class are also included all children with edema. This type of classification has been widely used; it has the merit of simplicity and gives information as to the actual nutritional status. The age of children must be known with a fair degree of precision. The method tends to over-estimate the number of malnourished children by not taking into consideration the height. Children classified as malnourished because they are underweight may be shorter than normal for their age with the result that their weight for height is normal.

A number of community surveys have been conducted using this type of classification. Bengoa (1) reviewed the literature in 1973 and found that in sample groups of approximately 1,000 children each to total 173,000, the prevalence of severe forms of PCM was on the average 2.6% in developing countries and the prevalence of moderate forms, 18.9%. In absolute figures, approximately 100 million children of preschool age would suffer from severe or moderate malnutrition.

The problem of overestimation is of little concern when one compares prevalence figures calculated by this method for different age classes. Such an analysis may help to identify the problem areas for future action programs. In Central America, for example, the second half of the first year appears to be the most critical period in El Salvador, while in Guatemala malnutrition is highly prevalent from six months up to four years.

<u>Height for age</u>. This indicator does not necessarily reflect the present nutritional status but estimates past and/or chronic malnutrition. It is seldom present in infant malnutrition and when observed is more likely to be the result of a small size at birth. Malnutrition must be prolonged over a long time before height is affected significantly and even then the deficit in terms of percentage by comparison to a reference child is much smaller than for weight.

<u>Weight for height</u>. The shortcomings of measuring height or weight for age can be overcome by relating weight to height, i.e., by comparing the weight to that of a reference child of the same height rather than of the same age. The comparison can be expressed in terms of percentage. It seems that 80% of the reference weight may be an acceptable cut-off point between adequate nutrition and malnutrition. By this method, the children who are stunted but are otherwise well nourished are no longer identified as being malnourished, and as a result, the number of malnourished children identified in a given population is usually smaller than when measuring weight for age and using a Gomez-type classification. This method is almost age-independent, bypassing, thus, one of the main difficulties of weight or height for age (2).

Waterlow (2) has proposed combining the information on weight for height with the growth performance of the child as provided by the parameter, height/age. A two by two tabulation can be prepared in which children are grouped vertically by percent weight for height and horizontally by percent height for age. The cut off points may vary according to the objectives pursued; convenient limits are usually 80% for weight for height and 90% for height for age; the inclusion of an upper limit such as, for instance, 120% for weight for height allows also for the assessment of the prevalence of obesity in the same population.

Action programs to remedy the situation could be suggested as follows:

Category I Long-term programs related to socio-economic development

Category II	None
Category III	Nutrition education
Category IV	Emergency rehabilitation
Category V	None
Category VI	Nutrition education

Arm circumference. Measurement of the arm circumference has been widely used as an index of malnutrition, either alone or in combination with height in the Quac stick test.

There is little increase in the arm circumference between one and four years and this measurement is, therefore, relatively but not totally independent of age during that period.

This is an increase of about 1 cm in three years, or 2% per year. By comparison, height increases by 8 to 9% per year and weight by 10 to 12% per year during the same period.

Shakir (4) has recommended the mid-upper arm circumference because it reflects the body mass changes of malnutrition and also because it is relatively constant between one and five years in well nourished children. Considering 16.5 cm as a reference figure during that period, one may draw a cut-off point at 75%, i.e., 12.5 cm as the limit under which the child is in a critical stage and requires immediate attention. Mild-moderate malnutrition will be represented by those having an arm circumference of 12.5 - 14.0 cm (76 to 85% of the reference). Anything above that figure and below 16.5 cm may be considered normal. However, the narrow distribution of measurements in a child's population makes it a rather insensitive index. This, added to the fact that the measurement usually lacks accuracy, appears to diminish considerably its significance.

The Quac stick (5) has been used in emergency situations, for its simplicity and the fact that it does not require elaborate or heavy equipment to carry around. It consists simply of a height measuring stick which is marked off in arm circumference measurements instead of height measurements. The values for 85% and 80% of the expected arm circumference for a specific height are marked directly at the corresponding height levels. If the child's actual height is below the level corresponding to his arm circumference, the arm measurement is greater than 85% of the arm circumference of an average child of his height. Therefore, he is not malnourished. Conversely, if he is taller than the level corresponding to the arm circumference marked on the stick, he is malnourished. According to reports from Bangladesh the Quac stick can be used efficiently under field conditions for prevalence surveys, but its value for screening purposes is doubtful. The method is simple and there is no need to know the age of the child; it requires, however, two measurements, the validity of which, in field conditions, is questionable. This is especially true of the arm circumference measurement.

Other measurements and indices, such as the ratios of arm circumference to head circumference or of chest circumference to head circumference, have been proposed but so far the data are insufficient to justify their use since they fail to show any advantage over more simple measurements. The same can be said of a number of other indices using weight and height, such as the indices of Quetelet, Kaup, Rohrer, etc. (2).

In conclusion, it appears that at present there are three indices that offer good prospects for increased use for nutrition surveillance and screening of malnutrition as it is usually observed in children, i.e., as a complex array of signs and symptoms caused by a combination of protein-energy insufficiency and repeated episodes of infectious diseases. These are weight for age, weight for height, and weight for height combined with height for age. The Quac stick and the arm circumference measurement appear to be useful tools in emergency situations. In normal circumstances, the classification of children according to their weight for height and height for age as proposed by Waterlow (2) appears to be the most promising in terms of reliability and usefulness of the information.

Biochemistry

A number of biochemical tests related to changes in the composition of plasma or urine have been proposed. They include the determination in the serum of the concentration of total protein and albumin and of the amino acid ratio; in the urine of the urea-creatinine ratio, the sulfur-creatinine ratio, the hydroxyproline index and the creatinine-height index. All these tests have in common that their applicability in field surveys is low and their results often vary from one geographical area to another. At the present time they should, therefore, not be considered for use as indicators of PCM in the community.

Vital Statistics - Age-Specific Death Rates

One of the characteristics of underdevelopment is the high mortality of childhood; the responsible factors differ with age. It is, therefore, useful to distinguish deaths among infants apart from those of children one to four years old, and also neonatal deaths from those in the postneonatal months. Postneonatal deaths are largely the result of acquired infections, but malnutrition may contribute significantly. In developing countries, neonatal deaths usually represent 40% of infant deaths; in developed countries they may constitute as much as 70%. A change in the ratio of deaths in the neonatal to those in the postneonatal period may, therefore, serve as a practical means to measure accomplishment in prevention of infection and malnutrition in infancy.

In developing countries, malnutrition and infection continue to be important causes of mortality at ages one to four years. In this age group differences between developed and developing countries are great, ratios between early childhood death rates being usually 1:10 and in some cases as much as 1:30. Death rates in this age group have sometimes been used as a measure of nutritional status of the population, although the importance of environmental sanitation is not to be underestimated.

A recent study (6) which was conducted in 13 areas of Latin America and of the Caribbean islands indicates that malnutrition is responsible for a high proportion of deaths of young children in that part of the world. Although conditions vary from one area to another, a similar pattern in the causes of mortality emerges. Among 7,318 deaths of children one to four years of age included in the study, malnutrition was found to be the underlying cause of death in 9% (range 0-18.4) of the cases and the associated cause in 48.4% (range 0-61.0). As a whole, malnutrition was directly or indirectly responsible for the death of children one to four years of age in 57.4% of the cases. There is no reason to suspect that the situation is much different in the areas not included in the study.

The ratio of death rates at ages one to four years to death rates in infancy has been suggested as an index of community nutritional status (7). This index, however, is not applicable to countries where nutritional status is good.

More recently the importance of death rates during the second year of life has been stressed (8). The largest proportion of deaths at ages one to four years occurs in the second year. The main causes of death in the less developed countries during this period are those characteristic of postneonatal infancy, when malnutrition assumes distinct importance. The second-year death rate has, therefore, been proposed as a practical index of community malnutrition. In a sense it is a better index of general social and environmental health than infant mortality, for it is less diluted with extraneous factors.

Death rates obviously are influenced by a number of interrelated factors not to be studied independently. Observations restricted to single factors believed to be causative can be misleading. Several attempts to correlate state of nutrition with death rates in early childhood have proved informative although not specifically indicative. Various other factors, such as infection, play an important role in mortality.

These indices, based on vital statistics, can, therefore, give some information on the probable importance of PCM in a community, but it will be almost impossible to build specific intervention programs on this basis unless they are comprehensive and cover a range of activities including nutritional ones, immunization and sanitation to mention only a few.

Birth Weight

An association between the weights of children at birth and the nutritional status of the mothers has been well documented in various parts of the world (9). A decrease of birth weight has been observed during times of famine, such as in the Netherlands in 1945. In countries where malnutrition is highly prevalent, a large proportion of children are born with weights under 2.5 kg. There is also recent evidence that supplementary feeding during pregnancy, providing as little as 100 kcal per day, will increase the birth weight of the offspring (10). Birth weight is influenced by a number of other factors such as socioeconomic status of the family, parity, stature of parents and grandparents, and infectious episodes during pregnancy. It is, nevertheless, a good indicator of the nutritional status of the community and as such will give a rough estimate of the prevalence of malnutrition in pre-school age children.

Socioeconomic Development

Socioeconomic development and nutritional status usually improve together because the factors responsible for their change are to a large extent the same. These factors, such as increased agricultural production, a higher degree of literacy, and an improved level of sanitation, are well known, but not easily quantified. The United Nations Research Institute for Social Development has proposed a socioeconomic development index which correlates quite well with the level of malnutrition. There are, however, some exceptions: In Chile, for instance, where weaning tends to occur very early, marasmus is already observed during infancy (11).

AVITAMINOSIS A

The ocular signs of avitaminosis A are well known and include night blindness, conjunctival xerosis, Bitot's spot, corneal xerosis, corneal ulceration, keratomalacia, and corneal scars due to avitaminosis A. The fundus of the eye may also show some typical lesions. The clinical signs are preceded by some biochemical changes, more particularly, a decreased level of vitamin A in the plasma and in the liver. The level of plasma vitamin A can be interpreted as high when over 50 mcg/100 ml, normal between 20 and 50 mcg/100 ml, low between 10 and 20 mcg/100 ml, and deficient when it is below 10 mcg/100 ml. A level of less than 10 mcg/100 ml is almost always associated with low liver reserves of vitamin A and with an increased prevalence of clinical signs of xerophthalmia. The low category may be caused by a low vitamin A intake but also by other factors such as an inadequate protein intake, a parasitic infestation, or a liver disease. As far as the vitamin A level in the liver is concerned, 20 mcg/g in a child and 10 mcg/g in an adult appear sufficient to offer protection for a period of 100 days, which seems reasonable. Xerophthalmia is closely associated with very low liver reserves.

The clinical diagnosis of avitaminosis A presents no major difficulty, but some criteria must be selected to determine whether xerophthalmia and avitaminosis A constitute a significant public health problem in a community (12).

Some suggested criteria, applicable to children below five years, are 1) Bitot's spot with conjunctive xerosis (X1B), 2) corneal xerosis + corneal ulceration with xerosis + keratomalacia (X2 + X3A + X3B), and 3) corneal scars (XS) (attributable to vitamin A deficiency). The presence of one or more of the three clinical criteria can be considered as indicative of a significant xerophthalmia problem warranting the emergency initiation of a control program such as the periodical administration of a 200,000 I.U. dose of vitamin A by mouth to every child under four years. On the other hand, biochemical data alone are only indicative of avitominosis A but not necessarily of xerophthalmia. Such evidence would justify the initiation of a vitamin A fortification program whenever conditions are suitable. The low cost of such a program is in keeping with the relatively low priority of the problem.

A few examples of recent surveys and of the programs of action undertaken in several countries are interesting in this context:

-- In the Philippines, a recent survey (13, 14) of Cebu Island has shown that of 1,715 children examined, 17% had serum vitamin A levels below 10 mcg/100 ml and 0.7% had xerosis of the cornea or even more serious manifestations of xerophthalmia. Such data would warrant immediate initiation of emergency measures.

-- In Bangladesh, where a national program of periodical administration of 200,000 I.U. of vitamin A has been launched, a survey conducted before the program was initiated indicated that 0.18% of the children had corneal scars probably due to a vitamin A deficiency.

-- In Central America, low serum values for vitamin A were found without evidence of xerophthalmia except in El Salvador, where the eye lesions were prevalent. A campaign of administration of large doses of vitamin A to young children was conducted in this country while a program of fortification of sugar with vitamin A is planned in Costa Rica and in Guatemala.

It appears, thus, that in the case of protein-energy malnutrition and avitaminosis A, the selection of indicators is a useful tool for the assessment of the prevalence and also for the evaluation of any action programs that may be initiated toward their control. This applies also to other deficiency diseases.

REFERENCES

1. Bengoa JM: The state of world nutrition. In Recheigl M (ed): Man, Food and Nutrition. Cleveland, Ohio, CRC Press, 1973.

2. Waterlow JC, Rutishauser IHE: Malnutrition in man. In: Early Malnutrition and Mental Development. Symposia of the Swedish Nutrition Foundation. Uppsala, Almqvist & Wiksell, 1974.

3. Burgess HJL, Burgess AP: The arm circumference as a public health index of protein-calorie malnutrition of early childhood (II) A modified standard for mid-upper arm circumference in young children. J Trop Paed 15:189, 1969.

4. Shakir A: The surveillance of protein-calorie malnutrition by simple and economical means. Environ Child Health 21:2,69, 1975.

5. Arnhold R: The arm circumference as a public health index of protein-calorie malnutrition of early childhood (XVII) the Quac stick: a field measure used by the Quacker Service Team in Nigeria. J Trop Paediat 15:243.

6. Puffer RR, Serrano CV: Patterns of Mortality in Childhood. PAHO/WHO Scient Publ No. 262.

7. Wills VG, Waterlow JC: The death rate in the age group 1-4 years as an index of malnutrition. J Trop Paediat 3:167.

8. Gordon JE, Wyon JB, Ascoli W: The second year death rate in less developed countries. Amer J Med Sci 254-357.

9. Rosa FW, Turshen M: Fetal Nutrition. Bull Wld Hlth Org, 43:785.

10. Lechtig A _et al_: Influencia de la nutricion materna sobre el crecimiento fetal en poplaciones de Guatemala. II Supplementacion Alimentaria. Arch Lationameri 22:117.

11. Joint Expert Committee on Nutrition: Eighth Report, 1971, Food Fortification. Protein Calorie Malnutrition. WHO Tech Rep Ser, 447.

12. Report of WHO meeting: Vitamin A deficiency and xerophthalmia. Tech Rep Ser No. (in press), 1975.

13. Solon FS _et al_: Vitamin A deficiency in the Philippines: a study of xerophthalmia in Cebu. Am J Clin Nutr (in press), 1975.

14. Solon FS _et al_: Vitamin A deficiency prevalence, causes and intervention in Cebu, Philippines. Xerophthalmia Club Bull 7:1.

ATTACKING THE MALNUTRITION PROBLEM

James E. Austin, DBA

Harvard University Graduate
School of Business Administration
Boston, Massachusetts

THE ROOT CAUSES

Lurking behind this conference's all encompassing theme of Food, Society, and Man is the stark reality of over 500 million human beings suffering from the erosive effects of malnutrition. It is our collective responsibility as concerned social scientists to explore means of alleviating the malnutrition problem. However, before examining solutions one should diagnose the causal factors underlying the problem.

POVERTY

Fundamentally malnutrition is a problem of poverty; its roots are economic. The biological need to consume more is clear; the desire to eat more is present; but the money to buy the additional food is absent. It is a problem of insufficient effective demand: people are too poor to eat more. Malnutrition is simply a reflection of inequitable income distribution. World malnutrition is not basically a result of producing too little food; rather it stems from the skewed distribution pattern of the food produced. The rich eat more. This is true among nations as well as within nations. The average world daily per capita calorie intake was 2,386, only slightly below the recommended daily allowance of 2,450. However, the average is deceiving: the more developed nations consumed on the average 3,043 calories per capita per day while the less developed consumed only 2,097.

The national averages cloak similar skewing of food distribution within each country. For example, Brazil shows an average intake of 2,541, which exceeds the recommended allowance of 2,450. When one examines intakes by economic strata within Brazil, a strikingly different picture emerges, 43.5% of the population is calorie deficient.

Taking the income factor into account is critical in estimating the true magnitude of the global malnutrition problem. The United Nations Food and Agriculture Organization estimated that as of 1970, 57 out of 97 developing countries had deficits of food energy supplies and that 462 million people were malnourished. This means that one out of every six people in the world is undernourished. If one takes cognizance of the uneven income distribution and skewed food consumption, the figures almost double. Reutlinger and Selowsky, using extrapolated income distribution data for 28 countries and estimated calorie-income elasticity functions, calculated that 670 million people have calorie deficits greater than 200 calories per day and another 447 million have daily deficits of less than 200 calories. These figures reveal that approximately one billion people - three quarters of the population in developing countries - are calorie deficient. The problem is accentuated by the fact that a large proportion of these people are also protein deficient.

It is clear that we are dealing with a major world affliction and that malnutrition is inextricably entwined with the problems of poverty and insufficient effective demand. However, this is not to say that problems do not exist on the food production and supply side.

FOOD SUPPLY AND POPULATION GROWTH

Over the past decades total world food production has shown an upward trend, rising from an index of 77 in 1945 to 133 in 1973 (1963 = 100). Total output increased at approximately the same rate in both the developed and developing nations. In spite of these production gains the rapid growth in world population has put us on a treadmill; we have made only miniscule progress in raising per capita food availability. Global population is growing at a compound annual rate of 2.0% and the pressure is particularly acute in the developing nations, where the rate is 2.6%. If unchecked the world population by the year 2000 will double, reaching 6.5 billion. The erosion of food production gains by this burgeoning population has been most severe in the developing countries. Whereas total food production in the developing regions rose from an index of 77 in 1954 to 132 in 1973, the per capita indices rose only from 96 to 103.

These small per capita supply increases on top of the inequitable distribution of global and country nutrient resources exacerbate the malnutrition problem even further. If population growth cannot be checked or food supplies greatly expanded, food intake of the poorest and most nutritionally vulnerable groups will tend to deteriorate even further. The emergence of the "world food crisis" in 1972-73 revealed how delicate the world food-population balance is. In 1972 world grain production dropped only 2.9%, but world grain prices soared. Wheat doubled, going from $1.76 per bushel in July 1972 to $3.25 twelve months later (1). Corn was selling at $1.32 per bushel in October and was 80% higher (at $2.37) a year later (2). Prices continued to rise throughout 1973, then softened somewhat in 1974, but today remain substantially higher than three years earlier. Rice and soybeans have experienced similar price increases.

These price increases take their most serious toll on low-income countries dependent on grain imports and on the poorer consumers within those countries. The basic question on the supply side is whether these higher prices reflect only a temporary aberration on the supply and demand curves due to poor weather and bad crops, or whether they represent a more permanent situation of short supplies in the face of an ever-expanding population. Population control via family planning programs is a slow process and even if zero population growth could be attained in the immediate future, the current age structure of the female population would leave us with a built-in demographic momentum which would keep growth rates up for several generations. Thus, there will continue to be an increasing number of mouths to be fed and here, again, the burden on the poor is compounded because they are the ones with the largest families.

Can we produce enough then? This depends on how much more land we plant and how much we can harvest from those plantings. The United States is the world's major grain exporter. In the fifties and sixties the United States paid farmers not to produce and millions of acres lay idle. Today that unused capacity no longer exists. In response to the higher prices the approximately 50 million idle acres were brought under cultivation. Increased plantings will not come from the United States. In the rest of the world it has been estimated that less than half of two billion hectares of arable land are actually cultivated (3). The constraint to exploiting this unused land, however, is economic: to bring on stream most of this land would require inordinate amounts of capital investment to build rural infrastructure. The Food and Agriculture Organization of the United Nations (FAO) stated that only about six million hectares of additional land (a 1/2% area increase) could be made arable quickly and then only at a cost of about $225 per hectare or a total of $1.3 billion (4).

Undoubtedly more land can and will be brought under cultivation but the increases in supply needed will more likely have to come from higher yields per unit of land planted. The countries with the greatest food needs and most acute malnutrition problems tend to have the least productive agricultural sectors. The developing countries in 1972 produced on the average 1.3 metric tons of grain per hectare; this is about 58% less than the average yields in the more developed nations. The poor both produce less and eat less. The key question is: can they produce more? Can yields be increased? The potential for significant improvement of productivity exists. High-yielding varieties (HYV) of rice and wheat were rapidly adopted in the late sixties. These seeds and the accompanying use of agrochemicals along with favorable weather gave rise to significant production gains, especially in India and Pakistan, where most of the HYV plantings have occurred.[1] The production shortfalls in the seventies do

---

[1]It should be noted that even when production increases are achieved, a significant portion (estimates run from 10% to 30%) of the harvest is lost due to inadequate storage and handling.

not portend the wilting of this "Green Revolution," but have revealed that genetic breeding will not give us a panacea. Higher oil prices and, therefore, fertilizer costs, along with limitations on irrigation resources and farmer resistance to adopt new technologies restrict the expansion potential of the "miracle seeds." Research will give rise to further genetic improvements and wider use of the high yielding varieties but the rapid gains have already been made. Progress will come, but slowly.

These constraints on both area and yield increases, along with the inevitable population increases, will create continued pressure on global food supplies. Perhaps we can avoid the spectre of "food triage" and the decisions of who shall be fed and who shall starve (5). Nonetheless, the heavy hand of Thomas Malthus still rests on our shoulders and his shadow casts an added darkness over the problem of malnutrition.

THE HEALTH-NUTRITION INTERFACE

Although malnutrition occurs with inadequate food intake, nutritional status is inextricably tied to other health factors. Food ingestion is in part dependent on the individuals having an adequate appetite. A sick child frequently has no desire to eat. Even when the food is consumed, its nutritional value is not realized unless the nutrients are absorbed by the body. Absorption is hindered if the child is suffering from diarrheal diseases. Conversely a malnourished child is less resistant to the onslaught of childhood diseases. The effects of a disease such as measles on a nutritionally healthy child are not serious, but can be fatal if the child is malnourished.

Thus, malnutrition exacerbates health problems and vice versa. The scarcity of health resources and sanitation infrastructure is particularly acute in the crowded urban slum areas and poor rural zones of most developing countries, thereby making the malnutrition problem even more tenacious.

KNOWLEDGE AND BELIEFS

Another set of factors contributing to the malnutrition problem is inadequate knowledge and adverse folk beliefs. In many instances people do not make optimal nutritional use of existing food resources because they are ignorant of different ways to prepare foods, or better means of storage, or even new, more nutritious yet inexpensive foods. Closely related to the knowledge problem is the holding of certain beliefs which lead to nutritionally negative effects. For example, in Malaya fish are not fed to children because they are erroneously believed to cause worms; thus, a potentially rich and available protein resource is under-exploited (6). Sometimes beliefs lead to over-exploitation of a poor nutrition food; in East Africa matoke, steamed plantain, which is a low protein food, is perceived to be a "cultural superfood" and is the staple food in the diet (7). Another custom, common in many Latin American countries, is that food is withheld from a child with diarrhea. This reduces protein intake and can cause the child to slip into a highly vulnerable nutritional position.

It is important to recognize that many of these beliefs are rooted in rational reasons given the constraints of the environment. Even if irrational, cultural patterns are deeply ingrained and not readily changed. Their presence must be fully recognized in nutrition planning efforts.

NUTRITION PLANNING

The foregoing examination of the causes of the malnutrition problem reveals that its roots are economic but that it also carries sociological, psychological, physiological, and political dimensions. This multi-faceted causality dictates that the problem be approached in a multi-disciplinary fashion.

Given the dominant role of the economic factor as a determinant of malnutrition, a logical focal point for nutrition planning efforts would be identifying means to improve income levels. If countries were to undergo radical socioeconomic restructuring such that significant income redistribution occurred, it is likely that the problem of malnutrition would largely disappear. Although this might be desirable, it probably is not politically feasible and, therefore, is unlikely to happen in most countries.

Another general economic approach is to assume that the broad development programs and growth in the national economy will generate income improvement and hence better diets. Experience has revealed that in spite of large increases in countries' gross national products the problem of malnutrition persists. The Brazilian experience, for example, corroborates this phenomenon. The economic benefits of national growth do not seem to "trickle down" to the poorer segments of society either quickly or significantly. The gap between the rich and the poor has generally widened.

Given the magnitude of the nutritional deficits to be filled and assuming the absence of income distribution changes, it is unlikely that foreseeable growth rates would materialize which would improve income sufficiently to significantly reduce the incidence of malnutrition in the near future. Thus, the task of nutritional planning is to find ways to make an impact on the malnutrition problem more quickly than would otherwise occur in the normal process of a country's economic growth. We are seeking short-cuts which accelerate nutritional improvements.

Nutritional planning is fundamentally an attempt to systematically address a national problem so as to find the most effective solutions within the socioeconomic and political constraints facing the country. Without planning, a country runs the risk of implementing costly programs which produce minimal impact. To date few countries, including developed nations such as the United States, have clearly stated food and nutrition policies derived from systematic planning. Concern for nutrition has traditionally been low, and even when addressed it has often been a stepchild of health programs or a peripheral dimension of agricultural activities. Only in recent years have economists (who traditionally played the major role in most countries' national planning efforts) become interested in nutrition. This interest has derived in part from a growing perceived importance among economists of the development of a nation's human resources as part of its basic capital stock. Expenditures aimed at reducing malnutrition become viewed as capital investments rather than mere social welfare expenditures.

Increasingly economists, nutritionists, physicians, and other scientists have begun to work together to develop systematic approaches to the problem of malnutrition. Such interdisciplinary marriages are difficult to consummate and preserve (8). Nevertheless, progress has been made. Many nutritional planning models have been developed, ranging from the very simple to the highly complex (9, 10, 11, 12, 13). Although all these approaches cannot be elaborated and reviewed here, the basic elements in the planning process common to most of them can be set forth.

The first step is problem identification. This entails assessing the nutritional status of the population to determine: which deficiencies exist, who is suffering them, and how serious they are.

In addition to the which, who, and how, one also looks at the why. Identification of the causal factors underlying the problem of malnutrition is critical. Accurate diagnosis is a prerequisite to proper prescription. In order to take a preventive rather than just a curative approach one must determine causality. The interdisciplinary approach is particularly useful in this phase. This analysis must recognize the multiplicity of and the interrelationships among causes.

Not all causes are amenable to adjustment. Thus, one must locate possible points of intervention. Having decided where in the system to intervene, one must then decide what intervention to use. There are no prepackaged solutions; each must be tailored to the individual situation.

After the alternative interventions have been delineated, it is necessary to select the best ones. This selection is necessary because resources are scarce. Therefore, the logical criteria are lowest cost and highest impact. This implies the need to establish very clear and quantifiable objectives against which impact can be measured. It also suggests the need for establishing priorities among population groups to be reached by the nutritional interventions. This will generally lead to the specification of infants, preschoolers, and pregnant and lactating mothers within low income families as the priority target groups due to their nutritional vulnerability.

Once the alternatives have been selected, the interventions must be implemented. The planning process continues throughout this operational stage in the form of program evaluation. Impact and performance should be measured and fed back into the intervention design and selection process. Thus nutritional planning must be iterative and dynamic if it is going to produce effective nutritional interventions.

Nutritional Interventions

Although one cannot specify optimum interventions which would remain valid for all countries, it is possible to delineate the general types of interventions that are open to countries undertaking nutritional planning and discuss their objectives, forms, and possible limitations. The interventions that will be discussed are: general food subsidies, targeted food subsidies, fortification, genetic breeding, new protein products, and nutrition education.

General Food Subsidies

This type of intervention is aimed at the income constraint. It attempts to increase the purchasing power of existing incomes, thereby permitting, theoretically, more food to be acquired and consumed. If this higher intake occurs, presumably nutritional improvement would follow.

General subsidies can take several forms but most commonly they appear either as price supports to farmers or price controls for consumers. Farmer price support subsidies mean that the government pays the farmers a higher price for their crops than would have otherwise prevailed under free market conditions. These subsidies can be aimed at increasing farmer incomes and at stimulating higher supplies of basic staples, thereby keeping prices lower to consumers. Subsidized price controls entail the government's ensuring lower prices to the consumer than would have existed without governmental intervention. In some instances the government subsidizes basic staples both at the producer and the consumer ends of the food system.

This general subsidy approach is costly from a nutritional impact perspective. The government can end up subsidizing a larger block of farmers and most of the consumers of the basic staple whereas the target group for the nutritional intervention is only a small percentage of these general subsidized groups. For example, the target group of infants, preschoolers, and pregnant and lactating women might represent, say, 30% of the population. If their consumption were similarly proportional, the government would have to spend $10 worth of subsidy in order to get $3 worth of benefit to the target groups.

In addition, there is no assurance that the incremental purchasing power will be totally translated into food expenditures. Although food costs dominate poor people's budgets, a portion of the added income might go to nonfood items such as clothing, housewares, or other less essential expenditures. These nonnutritional leakages may mean that only $2 of the $10 will go to target group food consumption. Even this $2 worth of food may not lead to dietary improvement of the vulnerable groups because of intra-family food distribution patterns. In many cultures family feeding priority goes to the father and other wage earners; therefore, increased food might be largely consumed by nontarget group members of the family. Finally, the types of food consumed and, therefore, the nutritional quality of the diet would not necessarily improve, even though the quantity consumed did increase.

Targeted Food Subsidies

These subsidies also are basically vehicles for increasing purchasing power, but unlike general subsidies they attempt to reach target groups more precisely. If the objective is to maximize nutritional impact rather than just redistribute income, then targeted subsidies are more desirable.

Several methods are used for targeting. A common one is income criterion; only families with incomes below a certain level are eligible for subsidies. For example, this is the method employed to determine participant eligibility for the food stamp program in the United States and ration shops in India. In Mexico the national marketing organization, CONASUPO, locates its stores in low income rural and urban areas.

Such income targeting significantly reduces the extent of excess coverage and thus the cost of subsidizing non-target group individuals. Still, the problem of the increased purchasing power possibly being diverted to non-food items remains. With such diversion the subsidy substitutes for existing food expenditures rather than supplements them. In addition, the difficulties of intra-family food distribution patterns remain.

Some of these limitations can be reduced if the income criterion is supplemented by a nutritional risk criterion and the subsidies are restricted to foods primarily consumed by the target group individuals. For example, the Women, Infants, and Children (WIC) program in the United States provides coupons to individuals in these categories that are deemed to be nutritionally at risk. This risk status is determined either by clinical examination or by income levels. The approach is preventive rather than curative. One does not have to become malnourished to qualify. The coupons can only be used to purchase a predetermined food package consisting of infant formula, dairy products, cereal, and juice. It is likely that some of these products, such as the cereal, milk, and cheese would be diverted to non-target group members within the family; however, no diversion would likely occur with the infant formula. The government of Colombia is currently designing a WIC type intervention in which coupons would be used for selected nutrition products, one being an infant formula derived from a locally produced high protein product called "Bienestarina." Although some leakage and substitution will still occur, such subsidized targeting does enhance the prospects of achieving higher impact at a lower cost.

Fortification

Fortification interventions strive primarily to improve the nutritional quality of diets. The goal is to add key nutrients to fill certain gaps in existing food patterns. Fortification was cited at the World Food Conference as one of the types of interventions that should be considered by countries in their nutrition efforts (14).

The form fortification interventions usually take is the addition to vitamins and/or minerals and/or protein/calorie supplements to a commonly consumed staple. For example, salt has been iodized in many regions, such as Central America, and has frequently led to a dramatic reduction in the incidence of goiter. Sugar was recently fortified with Vitamin A in Costa Rica and Guatemala. Rice, wheat, and corn have been fortified with amino acids and protein supplements in longitudinal research projects in Thailand, Tunisia, and Guatemala, respectively. The carrier of the fortificant is usually sought to be one with high consumer acceptability, universal consumption, and reasonably centralized processing.

Fortification is particularly useful where the nutritional deficiency is common to most of the population. The less broad the deficiency, the greater is the excess coverage. Because the carrier is consumed by the majority of the population, nonnutritionally needy groups would also be consuming the fortificant. The cost of this extra coverage is not so serious with vitamins and minerals; for example, fortificant costs of vitamin A in Guatemala are only about 1/10 of a cent for each pound of sugar fortified. However, protein fortification is much more expensive, running perhaps around five cents per pound of food fortified. This means that the economic burden of not being able to target it is quite significant.

Because fortification is an "invisible" product improvement, it is difficult to get consumers to pay the incremental costs, if they have a choice. Government subsidies may be imperative, and, therefore, cost-effectiveness remains a critical criterion in weighing this intervention.

It should also be noted that the impact of protein fortification will be severely limited if the target groups are suffering from calorie as well as protein deficiencies. The protein would be burned up for energy purposes rather than being available for body building. Protein fortification is a costly way to solve a calorie problem.

Genetic Breeding

Genetic breeding is fortification by means of biology rather than food technology. The objective is the same as the previous intervention: improvement of the nutritional quality of the diet.

Nutritional improvement via genetics has focused primarily on the cereal grains. Cereals account for 52% of the average world daily per capita intake of calories and 47% of the protein intake (15). The figures are even higher in the low income countries. Generally, cereal grains are nutritionally poor in lysine, threonine, or tryptophan. High lysine varieties of maize have been developed in the form of the opaque-2 and floury-2 mutants. Similar high lysine lines have been identified for sorghum, and a barley line, called Hiproly, has been found which was about a 25% increase in lysine content. Similar advances have been made with rice and wheat.

Genetic breeding shares the nutritional impact limitations of fortification via food technology, but it does have the advantage of eliminating the administrative and logistical complications inherent in adding a fortificant. On the other hand, farmers have to be convinced to use the new seeds and this entails the problems inherent in new technology adoption and farm input supply provision. Some nutritionally improved varieties have not retained other seed characteristics important to the farmer, e.g., yields and disease resistance. In the absence of price incentives for superior nutritional quality, the quantitative yield potential will continue to dominate the farmers' seed selection decision process.

New Protein Foods

Considerable effort has been placed on the development of high protein products aimed at nutritionally vulnerable groups. The goal is basically to improve the quality of food consumed by the target individuals.

A primary criterion in the development of these products has been economic. They are supposed to be low cost in order to be within the economic grasp of the poorer segments. They also are designed to carry a higher protein content and quality than traditional foods; most of these new products have a protein content around 25%. Attempts have also been made to base these products on locally available protein resources, especially where those nutritional resources had previously not been exploited for human consumption. Incaparina, based on cottonseed meal, was developed by the Institute of Nutrition for Central America and Panama (INCAP) in Guatemala. Kaset Protein, based on mung bean protein which had previously been a discarded by-product in the production of starch noodles, was designed by the Institute of Food Research and Product Development at Kasetsart University in Thailand. The Colombian Institute of Family Welfare has developed the soy-based Bienestarina for use in its national nutrition programs.

These products hold the significant potential of nutritional vehicles tailored to the needs of key at-risk groups. Nonetheless, only about half of the approximately 60 such products developed around the world have been successfully launched and remain on-going (16). Of all the interventions, new protein products perhaps have encountered the most barriers: taste, texture, cookability, poor distribution, costly packaging, limited government support, and excessively high prices. For some of these new products most of these barriers have been overcome, but they are difficult ones to surmount. The potential of this intervention is high, but the realization of that potential will come only with considerable effort. It should also be mentioned that the nutritional impact of protein products, like protein fortification, depends on first filling the calorie gap.

Nutrition Education

A final commonly cited and used intervention is nutrition education. Efforts here are attacking the knowledge and belief barriers. The aim is to change food behavior so that better use can be made of existing or new nutrition resources.

Nutrition education has traditionally taken the form of face-to-face communication. The educator is communicating with individuals or small groups on a personal basis. The other basic alternative is to use mass media such as radio, television, newspapers, and posters. The form of presentation and the content of the messages conveyed are critical determinants of the effectiveness of either basic approach.

Proponents of the face-to-face approach cite its superiority as an attitude change and its greater flexibility due to the personal interaction mode. Those who favor mass media point to the greater coverage possible on an economical basis and the experience of commercial advertising as an effective communications tool. In fact, both approaches are relevant and should be viewed as complementary and reinforcing. Nonetheless, behavioral patterns change slowly, especially when people face severe resource constraints. Consequently, nutrition education will not likely produce rapid impact.

AN INTEGRATED APPROACH

Having delineated these alternative interventions, we once again face the task of having to choose among them. All developing countries operate under severe resource constraints. Thus, it is essential to take the previously mentioned cost minimization - impact maximization approach to intervention selection.

However, in making these choices, we should remember that interventions are not mutually exclusive. In fact, the optimal approach is to employ a portfolio concept in designing a national nutrition strategy. We need to look for ways to combine interventions in a synergistic manner so that we achieve economies of operation and complementarity of impact. This portfolio concept also dictates that the relationship of traditional nutritional interventions to other interventions be considered. Experience has shown that major inroads on the problem of malnutrition can be made by introducing potable water, public sanitation measures, and inoculations. The multifaceted causality of malnutrition means that a broadly integrated approach holds the greatest potential for impact. Different interventions will generate their impacts at different points along the time spectrum. The short, medium, and long run impact dimensions need to be blended together in a complementary fashion. The intervention mix must remain flexible. It must be adjusted to reflect operational problems and progress as well as the ever changing dynamics of the national and global food system.

Currently, major national nutrition planning is being undertaken in Brazil, Indonesia, Colombia, the Philippines, India, and Costa Rica. Incipient efforts are underway in several other countries. Global consciousness has been raised. An attack is being made on the extremely large and tenacious problem of malnutrition. Systematic nutritional planning will increase the chances of these efforts to make an impact on the problem, but it should be once again recognized that malnutrition is fundamentally a problem of poverty. Nutritional interventions can be viewed, in one sense, as cost-effective and politically palatable means of income redistribution. Consequently, we should realize that fundamentally, nutritional plans will be meaningless unless they are backed by a solid political commitment to attack the problem of malnutrition. Such commitment, unfortunately, is often the scarcest resource of all.

REFERENCES

1. United States Department of Agriculture: Wheat situation. C.I.F. Prices, Rotterdam. Washington, D. C., Economic Research Services, Aug, 1975.

2. ___________: Feed situation, Chicago Cash Prices. Washington, D. C., Economic Research Services, Sept, 1975.

3. Blakeslee LL, Heady EO, Framingham CF: World Food Production, Demand, and Trade. Ames, Iowa, Iowa State University Center for Agricultural and Rural Development, 1973.

4. United Nations: The World Food Problem - Proposals for National and International Actions. Rome, 1974.

5. Paddock W, Paddock P: Famine-1975. 1967.

6. Jelliffe DB (ed): Child Health in the Tropics. London, Arnold Publishing Company, 1968.

7. ___________: Child Nutrition in Developing Countries. Washington, D. C., United States Agency for International Development, 1969.

8. Dwyer J, Mayer J: Beyond economics and nutrition: the complex basis of food policy. Science 188: May 9, 1975.

9. Joy L: Food and nutrition planning. J Agri Econ 24: Jan 1973.

10. Levinson FJ, Call DL: in Berg A, Scrimshaw NS, Call DL (eds): Nutrition, National Development and Planning. Cambridge, MIT Press, 1973.

11. Cooke T, Pines J: Planning Nutrition Programs: A Suggested Approach. Washington, D. C., Office of Nutrition, Agency for International Development, 1973.

12. Berg A, Muscat R: The Nutrition Factor: Its Role in National Development. Washington, D. C., The Brookings Institution, 1973.

13. Perissi J, Sizaret F, Francois P: FAO Nutritional Newsletter, No. 3, 1969.

14. United States Department of Agriculture: The world food situation and prospects to 1985. Foreign Agricultural Report, No. 98. Washington, D. C., Economic Research Services, Dec 1974.

15. Food and Agriculture Organization: Agricultural Commodity Projections - 1970 to 1980. Rome, United Nations FAO, 1972.

16. Orr E: The Use of Protein-rich Foods for the Relief of Malnutrition in Developing Countries: An Analysis of Experience. Tropical Products Institute, Aug 1972.

FOOD AND SOCIOECONOMIC DEVELOPMENT

Cipriano A. Canosa, M. D.

"La Fe" Children's Hospital
Valencia, Spain

Food and health are two inseparable fundamental biological principles on which the evolution and progress of mankind are partially based. Health is the final consequence of many interrelated and interdependent physical, biological, and sociocultural factors, among which food and, therefore, nutrition is one of the most important (1). Furthermore, the timing between food and age is of critical importance to insure full development of the genetic potential and to maintain good health (2). Quantity and quality of nutrients in early life have a direct effect on physical growth, biochemical maturation, body composition, physical fitness, work capacity, possibly mental development, and, of course, morbidity and mortality (2, 4, 14, 15, 16, 17, 18, 19, 20). However, in spite of the amount of work, knowledge, and progress made, there is no specific answer yet as to what is the best food for the right age to obtain and maintain perfect health.

Hunger, the world's number one health problem, is due to multifactorial interdependent ecological factors. Among the most important of these are:

-- lack of food: production, distribution, storage, availability, and consumption

-- population explosion, which occurs mainly among developing societies which simultaneously have low food production and low per capita food consumption

-- sociocultural factors: mode, migration, acculturation, poverty, ignorance, taboos, biases, faddism feeding habits, etc.

Each of these three major factors is determinant of poor nutrition and mainly affects two at-risk groups: pregnant and lactating women and infant and preschool children (3, 4). In developed societies poor nutrition, if present, is most likely to be due to intrauterine and sociocultural factors, since lack of food and population explosion are, by definition, unimportant (5).

SOCIOCULTURAL FACTORS

Mode, Time, and Place

In primitive societies mode, time, and place play relevant and decisive roles in food patterns and behavior and, therefore, act on the nutritional status (6). Meals are an important daily social family event. The usual female preparation, the ritual of food distribution, the priorities according to sex and age, and the intensive and continuous social interactions are some of the sociological components of feeding in these societies. Important social events (weddings, birthdays, mournings, etc.) are celebrated with food festivities with expenditure often of a significant share of the family income (6).

To the contrary, in progressive and aggressive industrial societies the family seems seldom to gather to enjoy meals together. Eating outside the family environment (brunches, snacks, prepared frozen foods, cold dishes, sandwiches, sweets, soft drinks, etc.) by all family members, irregularly and throughout the day, has become a frequent pattern of food behavior. Festivities, as well, are celebrated differently. Food is being rapidly replaced by drinking parties, which are definitely responsible for some of today's nutritional imbalances. This distinctive behavior by families relative to food intake, among other factors, has significant nutritional implications, being associated with protein-calorie malnutrition (PCM) in developing societies and with poor nutrition as well as obesity in high risk groups in the developed ones.

MIGRATIONS

Mass migration of large populations for political, economic, and social causes is an old phenomenon still prevalent in some areas today. These "voluntary-forced" migrations change dramatically numerous social and economic patterns of societies whose cultures are old and have deep roots. Food behaviors and feeding practices are among these.

Recent examples are to be found in the periodic moving of Mexican American families north from Mexico and the southern states mainly to Colorado and California to work in crop harvests. These already malnourished populations suffer further deprivation during mass migrations when their fragile, borderline states of nutrition are aggravated. New and acute cases of kwashiorkor and marasmus often result (7). Government nutritional aid and nutritional education have not produced significant differences between control groups and experimental groups receiving nutritional education, even with ten scheduled, carefully planned visits from specially trained social-nutritional workers (7).

Recently in Europe mainly male Italians, Greeks, Spaniards, Yugoslavs, Turks, and Portuguese have moved *en masse* to work in factories in the northern continental countries of France, Germany, Holland, Belgium, and Switzerland. When the male was able to earn a good salary, the remaining members of the nuclear family moved north. One of the unwanted side effects of these migrations was the birth of a new subculture entrenched in those well-developed countries. Whole families and even whole small communities from the southern European countries relocated in different and hostile environments and were subjected to new physical, social, cultural, economic, and biological conditions. Undergrown towns emerged as a new reality: poor housing, poor sanitary conditions, inadequate schools, social isolation, a new and difficult language. All were further aggravated by the "saving obsession" to finance the return to their homelands with cash to start a new "third beginning." Pediatricians from Paris, Geneva, Hamburg, Amsterdam, and Brussels were suddenly faced with a kind of pediatric pathology which had practically disappeared from their countries: tuberculosis, repeated bouts of diarrhea and respiratory diseases, iron deficiency anemia, parasitic infestations, and severe cases of malnutrition were again seen regularly. New and foreign food patterns and behaviors were one of the factors responsible for this unhealthful situation. A few years later this low laboring group was able to raise its standard of living and adapt themselves to the previously hostile environment; but the phenomenon was then repeated with migrant North African laborers.

Internal migrations in all countries, mainly movement from rural to urban communities, is today a universal phenomenon. In the United States approximately 70 percent of the populations live in communities over 5,000 inhabitants. In Spain sixty percent, and in Africa and the Far East, 80 to 85 percent of people live in rural communities of less than 10,000 inhabitants. This permanent influx of families - and in some cases whole communities go from rural to urban centers - results in dramatic changes and the necessity to adapt to new sociocultural patterns among which are food practices and behaviors (8).

ACCULTURATION

Primitive societies are examples of changing feeding practices directly dependent upon sociocultural changes imported from alien cultures. Eskimo and subarctic societies in times past ate the products of game hunting stored in the world's largest natural deep freeze. Depending on the season, the eskimo ate both fish and mammals. With the introduction of the gun into his culture, his food behavior dramatically changed: he was able to kill more than he needed with greater ease, with a resultant decrease in both his primitive hunting skills and in his physical exercise (9).

The fur trade, however, fast depleted his game stock. Fur trading and storage facilities introduced a new way of life which had its effect on food patterns and behavior. Wheat flour, sugar, lard, tea, coffee, canned milk, and meat all became available and replaced the need to hunt and fish. Finally civilization brought him the luxury foods - cakes, sweets, crackers, and potato chips. Along with the change from high to lower animal protein diets, food distribution also changed with acculturation. Initially food was obtained by kinship rights, though food was never denied to a person in need, even outside the family environment. The trading post and fur trading changed all of this. Family kinship rights were questioned and eventually were lost: each family was then on its own to obtain its daily food. Paradoxically, poor Eskimo families could eat better than the more affluent ones, because they still ate fish and large game, wild plants, insects, small game, and eggs instead of the "luxury" foods.

When North American Indians and Latin Americans were introduced to powdered milk and wheat flour provided them through the United States Aid to International Development (USA-AID) programs, which were seeking to improve their nutrition, these people found the foods foreign to their tastes and their preparation foreign to their cultures.

Another example of the relation of feeding practices and acculturation was recently demonstrated in a study of Japanese natives living in Japan, Hawaii, and California (10). The feeding habits of more than 10,000 adults of Japanese ancestry were surveyed in their natural habitats. A dietary acculturation score demonstrated a range, according to the living locations, from low for persons living in Japan to very high for those living in California. There were marked differences in their dietary patterns and variations in their nutrient consumption. These dietary differences, among other factors, have had bearing on the higher incidence and prevalence of coronary heart disease in populations of Japanese origin living in the United States than in those genetically similar living in Japan.

EDUCATION

The educational level of a community or country is practically synonymous with its nutritional education. Clinical and biochemical nutritional indices in developed and developing societies are directly related to their sociocultural levels. The Ten State Nutritional Survey of American preschool children revealed marked deficiencies in serum iron and vitamins A and C when children of low income states were compared with children of high income states (11). Similarly, lower levels for hemoglobin, serum vitamin A and C, and riboflavin were found among poor whites, blacks, and Spanish-Americans.

Dietary intake of several essential nutrients in very poor rural Guatemalan communities, in which illiteracy rates were above 85 percent (11) demonstrated serious deficiencies among pregnant and lactating women mainly in total calories, animal proteins, vitamins A and C, and riboflavin. The placental composition in those undernourished women, in comparison with well nourished pregnant women from Iowa City, Iowa, demonstrated marked quantitative and qualitative differences. It is probable that these deficiencies were due to nutritional deficits suffered, though the genotype make up could also have been an accountable factor.

Sereni *et al*. (12) demonstrated that iron deficiency anemia occurred much more frequently in the children of the laboring class in urban Italian preschool children and correlated with the sociocultural status of their families.

Finally, Canosa *et al*. in Spain (5) demonstrated that taking height and weight as indices of nutritional status, 24.8 percent of all preschool children from six months through 72 months, admitted to "La Fe" Children's Hospital, were below the tenth percentile /Iowa/. The populations studied belonged to lower social class groups. The chief explanation for this high percentage of malnourished children was poor quantity and quality of nutrients along with repeated bouts of upper respiratory infections and gastroenterocolitis.

In summary, in both developing and developed societies, the nutritional status of a population is a direct consequence of the sociocultural and educational levels of their families. This is a worldwide, well known phenomenon.

FEEDING PRACTICES

Feeding practices, determined by a large variety of sociocultural and economic factors and many times difficult to identify and quantify, are fundamental in the definition of the status of nutrition of a population. In stable societies feeding practices are established early in life and are difficult to change. On the other hand, in fast developing communities sometimes these are more easily altered, mainly through mode, advertising, and better nutritional education and knowledge. Feeding practices for infants and during early life are directly established by their parents' ancient and deep rooted cultural practices. Therefore, except for countries with a high percentage of breast feeding habits, children who soon after birth are fed and nourished according to their parents' nutritional knowledge and education, may suffer from the limitations and ignorance inherent in these practices.

Dietary surveys on infant feeding practices were conducted in Spain in 1967 and 1971. A study of a randomized, selected sample of children from birth to eight months of age, chosen throughout the country according to standard statistical methodology and involving rural, semirural, and urban communities of 2,000 or more inhabitants, regardless of social class and economic conditions, was carried out. In 1967 the per capita income for the country was $760; in 1971 it was $1,300 (in United States dollars). A total of 10,623 children were studied in 1967 and 11,282 in 1971.

Using an adapted Warner scale, the social class percentage distribution was practically the same for 1967 and for 1971; thus, the differences which could be observed between the two samples could not be attributed to this variable. In 1967, 84 percent, and in 1971, 80 percent of the studied sample belonged to medium low and low social classes.

The type of feeding offered the infants and children was defined as follows: milk feeding, breast and/or bottle; mixed, milk and/or soft diet; and basic, soft and/or solid food. The distribution according to these three types of feeding in both 1967 and 1971 shows a tendency to increase the mixed feeding type in early life. No significant differences were found in relation to social class, mothers' ages, number of children, geographical distribution, or population size.

In the survey all parts of Spain were studied. The regions of Madrid, Catalonia, the Mediterranean area, and

Galicia are representative of the wide variety of regional customs and characteristics:

- -- Madrid, a culturally mixed community
- -- Catalonia, a prosperous and more socially homogenous sample
- -- The Mediterranean area, a rapidly developing autonomous region
- -- Galicia, a poor, isolated, backward region.

The relative decrease in breast feeding from 1967 through 1971 was more evident in the Catalonia and Mediterranean regions. No clear explanation is available to justify the lowest breast feeding in the poorest region, Galicia.

When one studies the breast feeding and social class figures, the decrease percentage is more evident in the lower class groups. No significant differences were found for breast feeding, social class, mothers' ages, and number of children. However, population size was directly correlated with breast feeding, showing higher incidence in smaller communities.

Breast feeding was replaced by commercial powdered milk products. Powdered milk products consumption vs. social class demonstrates a linear relationship between these variables. Mothers' ages were not correlated; however, the size of the population was highly correlated: the larger the population, the larger the powdered milk consumption.

Besides the socially determined feeding practices in open communities, medical and nutritional progress are also of great importance to large groups of newborns of low birth weights, either small or appropriate for their gestational ages, as well as for children of normal birth weights. A retrospective comparative study done at the same hospital among two groups of newborn children exposed to two different feeding methods was carried out in England (13). From 1961 through 1964 during the first days of life it was customary to reduce or to postpone oral formula feedings as well as significantly to reduce total food intake. There were observed differences in fluid intake for the two groups of newborns and, therefore, for essential nutrients, as well as in body temperature. For many years the great majority of newborn children hospitalized in nurseries throughout the world were fed in quantity significantly less than what today is considered appropriate; they were also kept at a lower temperature. Follow-up studies of the 1961-1964 sample revealed lower scores on their intellectual development. Probably some of these differences could be due to the consequences of deficient food intake and lower body temperature during the newborn period. This study also shows that feeding practices carried out at newborn nurseries have been modified, but these institutionalized nutritional feeding practices did expose a vast number of newborn children for long periods of time to significant lower nutrient intake during this critical period of life. The immediate and longterm effects on physical growth and possibly on mental development are now being investigated (13).

During preschool years children are still fully dependent for their nutrition on parents' feeding habits and nutritional behavior. A retrospective nutritional survey was carried out among 5,223 children from birth through 18 months of age, who were admitted, from 1971 through 1973, at Valencia, "La Fe" Children's Hospital. A simple, but detailed nutritional questionnaire was used to obtain information on their food patterns and behavior. Some of the outstanding findings were: up to one month of age, 12.3 percent of all newborns were breast fed; at three months, 7.3 percent; and at six months, 2.7 percent. At three months of age powdered milk formulas were used in 53.5 percent of the children; 35.9 percent were fed homogenized pasteurized cow's milk. During the first six months of life 85.6 percent of the children did not receive vitamin D, either directly or in milk formulas.

The incidence rate of rickets observed in 489 children at "La Fe" Children's Hospital from 1973-1975 was 25.54 if one considers such risk factors as illiteracy rate, population educational level, income per capita, and such meteorological conditions as sunny days and air pollution in Valencia. During this time the very high incidence of rickets can be explained only in terms of the limiting sociocultural factors of poor feeding habits and poor hygiene. Thus, rickets in this part of Spain must be considered as a sociocultural disease.

SUMMARY

A limited and schematic review of some of the most significant sociocultural factors related to food intake throughout life, both in developing and developed societies, is presented. It becomes apparent that in both societies, sociocultural factors are of fundamental importance in the determination of feeding practices: both excess and deficit are of great relevance to physical growth, and possibly mental development, as well as to morbidity and mortality.

It is well known that in developing societies low social class and poor education are synonymous with poor nutrition. However, when whole populations in developed societies are studied, it is proven that nutritional deficiencies still exist among marginal groups of these people, as a result of poor education, erroneous feeding practices, nutritional taboos related to health and disease, biases, and prejudices. Pediatricians and nutritionists are well aware of the numerous wrong concepts in feeding practices as well as the lack of nutritional knowledge in great sectors of society. It must follow that in these societies a relatively large group of children are poorly fed during the critical periods for growth and development because of sociocultural factors.

Only through the appropriate and enforced nutrition educational programs at all learning levels - primary through higher education - as well as better and more complete nutritional curricula in medical and public health schools can we perhaps solve the problem, or at least significantly alleviate it.

This is the hope.

REFERENCES

1. Canosa CA: Ecological approach to the problems of malnutrition. In, Scrimshaw NS, Gordon (eds): Malnutrition, Learning and Behavior. Cambridge, MIT Press, 1969.

2. Birch HG: Malnutrition, learning and intelligence. Am J Publ Health 62:773-784, 1972.

3. Committee on International Child Health: Malnutrition in the world's children. Pediatr 43:131, 1969.

4. Vahlquist B, Stapleton T, Behar M: Conclusions and recommendations of a workshop on nutritional problems. J Pediatr 80:163, 1972.

5. Canosa CA, Martins FJ, Roques V, Folch F: Malnutrition before and after birth in Valencia. Mod Probl Pediatr 14:20-37. Basel, Karger, 1975.

6. Cravioto J, Delicardie ER: Ecology of malnutrition - environmental variables associated with clinical severe malnutrition. Mod Prob Pediatr 14:157-166. Basel, Karger, 1975.

7. Chase PH et al: Effectiveness of nutrition aides in migrant population. A J Clin Nutr 26:849-857, 1973.

8. Lewis O: Los Hijos de Sanches. Mexico, Mortiz, 1964.

9. Gonzales N: Changing dietary patterns of North American Indians. Nutrition, Growth and Development of North American Indians. Washington, D. C. Department of Health, Education, and Welfare Publication, 72-76.

10. Tillotson J., Kato H, Nichaman MZ et al: Epidemiology of coronary heart disease and stroke in Japanese men living in Japan, Hawaii, and California: methodology for comparison of diet.

11. Schaefer AE: Epidemiology of pre and postnatal malnutrition in the USA. Mod Probl Ped 14:9-19. Basel, Karger, 1975.

12. Sereni F, Marcheson C, Reali E, Principi N: Some aspects of child malnutrition in Italy. Mod Prob Pediatr 14:38-47. Basel, Karger, 1975.

13. Davies PA: Perinatal nutrition of infants of very low birth weight and their later progress. Mod Probl Pediatr 14:119-133. Basel, Karger, 1975.

14. Chase PH, Canosa CA, O'Brien D: Nutrition and biochemical maturation of the brain. Mod Probl Pediatr 14:110-118. Basel, Karger, 1975.

15. Gluck L, Silverman WA: Phagocytoses in premature infants. Pediatrics 20:951-957, 1957.

16. Graham G: Effect of infantile malnutrition on growth. Fed Proc 26:140, 1967.

17. McCracken GR jr, Eichenwald HF: Leukocyte function and the development of opsonic and complement activity in the neonate. Am J Dis Child 121:120-126, 1971.

18. Martinez M, Conde C. Ballabriga A: A chemical study on prenatal brain development in humans. Mod Probl Pediatr 14:100-109. Basel, Karger, 1975.

19. Viteri F: Physical fitness and anemia. Proceedings International Symposium on malnutrition and function of blood cells, Kyoto, 1972. Tokyo, Nat Inst Nutr:559-583, 1973.

20. Winnick M: Changes in nucleic acid and protein content of the human brain during growth. Pediat Res 2:352-355, 1968.

21. Berg T: Immunoglobulin levels in infants with low birth weights. Act Pediatr Scand 57:369-376, 1968.

22. Canosa CA: Placental biochemical modifications in malnourished mothers in rural Guatemalan areas. 12th International Congress of Pediatrics 1:105, 1968.

23. Dobbing J, Sands J: The quantitative growth and development of the human brain. Arch Dis Child 48:757-767, 1973.

ETHICAL AND CULTURAL ASPECTS IN HUMAN FOOD BEHAVIOR

IMPLICATION IN FOOD PLANNING

J. Trémolières, Ph.D., M. D.
G. Petrizzelli, Ph.D.

National Institute of Health
and Medical Research
Hôpital Bichat
Paris, France

NUTRITIONAL PLANNING

A concerned society must include the implicit social and cultural factors when consideration is given to the basic components of nutritional planning.

Nutritional planning in a given society, at a given time, means the determination of the amounts of different types of foods which are required to meet the needs of physical growth and development and health maintenance and to keep to a minimum the diseases of nutritional origin. This estimate should then become the intended goal of production, of technology, of commerce, and of the legal and monetary market regulations. This type of planning is only valid if the following three conditions are fulfilled:

-- a clear definition of food items, measures of their quantity and quality, and recognition of the factors that tend to modify them

-- a precise determination of diseases issuing from inadequate food consumption

-- the cause and effect relationship between the ingested nutrients and their consequence in health.

Nutritional planning is, in reality, an extension of the legislative power of the political society exercised in the field of consumption goods: a social consensus which entrusts to "experts" the responsibility to organize food production and distribution through an objective and logical technique called planning. The responsibility of food supply is, in fact, the keystone of a society.

For optimal food planning a society may need to make certain agricultural changes. For example, planting fields of corn may be a better choice than planting wheat; and small multicrop farming may have to give way to intensive cultivation of specific crops with property, farming techniques, and the price structures being modified as a consequence.

Today food processing can be categorized on three levels:

-- In underdeveloped countries it is often primitive, following the cultural pattern of a society.

-- It is artisanal, as it is practiced in rural areas and small communities where housewives' habits, their limited kitchen facilities, and the social significance of meals all have bearing on the process and the incidental service rendered.

-- It is an agglomerate industry that is found in large urban areas.

In artisanal processing raw products of the land are transformed into bread, cereals, cheese, and wine by simple, small-scale techniques; produce goes directly from the farm to the village markets; and fish, fowl, and cattle are killed and dressed for immediate local distribution and consumption.

Quite in contrast is the large, urban manufactory which converts the same grains, produce, fowl, fish, and cattle from the raw state through processing and packaging to foods of safe and regulated quality either for immediate consumption, for relatively long-term storage, and/or for transportation to populations both near and far for consumption.

On the village marketplace a man knew what he was buying and why. In today's industrial society the immensity of production precludes the consumer's knowing what he is really buying. The link between seller and buyer has been broken; even the object of the transaction looks, feels, and tastes different. In place of a person-to-person relationship there are a multitude of middlemen standing between the producer and the user, demanding to be paid their due wage, thus vastly increasing the price of the product. How can social and governmental structures determine regulations and pricing that are fair and equitable for those who produce raw foods for and those who buy prepared foods from such vastly different purveyors?

Food is not what it seems to be; disease is not what we say it is. We have come to a Babel of confusion quite like the Biblical one. It is only in finding the Word that man will be able to rebuild his city.

Planning and forecasting must not be regarded as magic mathematical calculations projected for the abstract average man. We hope to be able to show that a correct definition of food, needs, and diseases can only be obtained with a broad approximation on the time and space scales. Such a definition includes the subjective and ethical aspects which influence the objective aspects related both to individuals and to micro-societies. These are dynamic aspects which are the transforming forces of evolution.

Analyzing the data fed to the computers reveals that large-scale forecasting for a hypothetical average man does not take into account either individuals or micro-societies.

FOOD

Like white light, which is a combination of red, yellow, and blue, a food results from the integration of three basic components: the nutritional, the psychosensorial, and the symbolic. We have already developed this common-sense concept (1, 3). Pleasure and evocation through image are as essential as nutrients.

These three endogenous parameters making up food on an individual level are intimately mixed with the sociocultural component of food, i.e., the Mother Earth which integrates work, social communication, and even the sense of life itself.

Animal psychology work by Tinbergen and Lorenz has shown that general behavior integrates a whole of intuitive notions, which are interrelated and ordered, into a sequence.

The philosophical implication of these facts constitutes too large a universe to be summarized in a few lines. Parallel to the objective, scientific knowledge, there exists a sensitive knowledge and an intuitive, poetical knowledge.

Mother Earth

Integration in the ecological milieu can be typified by the relation that a baby establishes between his hunger and the face of the mother who satisfies it. The natural surroundings, the green or rural spaces, are reminiscent of the nourishing mother's image. One could probably give a psychoanalytical interpretation of landscapes.

Man does not enjoy eating alone; he wants to remain faithful to his family's habits, to savor the taste of country bread, to pursue the myth of natural food. All of these behavioral constants show that through the single act of eating man and his environment are integrated.

Do we not really want to suppress the hostile image of wilderness, desert, deep forests, or asphalt and concrete jungles and experience, instead, the peaceful picture of a village surrounded by fields or of a cottage with a wisp of smoke coming from its chimney? Man will always be attached to the Mother Earth image. If wine, bread, and cheese are derived from the earth, they also have the power to evoke it. The same cannot be said of a single cell protein steak or of chemically treated flour. In spite of the most advanced architecture and urbanization, the affluent citizen escapes to his country house and looks for the green spaces whenever possible.

Nutritional planning must take into account the transformation of the earth's surface, because, through his food, man is connected with it. Will it be possible to find a poetic meaning for additives or bio engineering? Or will it be better to hide them, as we do the body functions? Will man be able to humanize the technological context in which he lives but which he fears?

The Work

Beyond food there is not only the appealing face of the mother, but later also that of the father: purveyor, defender, builder of home and the village. Through him we come to understand the concept of work: work, which provides the nourishment, which cultivates the earth, which humanizes nature, remains present in the food. Aside from water, which ideologically remains pure, untouched, mineral, all nourishment is the fruit of man's work, the result of his art and his ability; and it is bound traditionally to a region and a culture. To eat bread, cheese, wine, and cake means to sanctify man's work which results from his culture. That culture combines making and tasting and living into a mysterious and undefinable essence. That "something" is an intuition, a poetic sense of goodness which creates the art of living.

There are foods which are the fruits of nutritional science and all the advanced technologies. They also have their adepts, their society.

There are also the convenience foods, the ones which we are obliged to eat because we take our meal at the cafeteria, or because we can only do our shopping at the supermarket on Saturday, or because we are not able to cook or do not have the time for it.

These are only a few examples of interaction of work upon food.

The Communications

The best way to feel the social climate of a group is to share a meal around the same table. Some examples:

-- solitary eating in the crowded company cafeteria

-- silent eating of the poor peasant working team

-- loquacious eating of the family happy to be gathered

-- high-tension eating of business meals, cocktail parties, receptions.

The table rituals create the emotional atmosphere and sketch the social context which will permit the words to be understood, even if they are only vague.

A nutritional planning which does not take into account the deep communicative meaning of a meal around the table will fall short of expectations because the nourishment will not correspond to the needs.

The Economy

The simplest mathematical translation of nutritional planning into practice can severely affect the socio-cultural factors of eating by creating economic strain.

International prices in a free-exchange system constitute a powerful constraint on the life of societies. The prices of soya, cereals, meat, and dairy products correspond to the production cost (plus profit), in countries which possess a thick humus, a favorable climate, and prior investments in mechanization; but they represent a shameful waste to the backward agricultures, generally handicapped by costly irrigations, hostile climate, and a scarcity of humus. These farmers are reduced to the role of proletarians and their production system is disrupted. The cost of technical evolution compels them to fall into line with monetary and commercial systems created by the industrial countries, which have accomplished their evolution decades ago.

The internationality of the markets brings about a concentration of structures favored by their credit organization, their technical expertise, and their marketing ability, which promote their establishment in large agglomerations, thus further affecting the ecological and ethnological systems.

Treaties, norms, and discriminating regulations are other barriers which eliminate products which do not strictly conform to the rules established by the industrial countries.

The dependence of the "Green Revolution" on complex social and economic factors is now clear.

Those factors, which we considered as fixed, have suddenly been upset and have brought into focus all the points that had been neglected. We now are painfully aware of the fact that fertilizers are partly a petroleum by-product, that tractors and plants run on oil, and that even the weather can let us down. The problem has been further complicated by the disappearance of most grain reserves, which were adroitly purchased by certain nations just before the trouble manifested itself.

This phenomenon has burst like fireworks upon us, and the stars keep falling. Sugar cane growers oppose beet culture; drought decreases by five percent the oil seeds crop; swollen stocks of strategic raw materials put pressure on prices. It is difficult to control the flow of sugar or cereals when prices depend on depleted stocks.

At this point of the analysis of the true definition of food we would have to face two questions:

-- What is really the goal of food planning? Does planning not require both a political and ethical choice? This question will be discussed at the end.

-- How do we improve our food data for planning?

Let us, therefore, leave to the "experts" the meditation that such facts require.

Fats, proteins, sugars, and starches are now raw materials, which have been stripped of their food characteristics and, therefore, can be defined rather simply by each type of society.

Their transformation and use bring about a myriad of foods which can be grouped into six or seven categories. The food belonging to each of these groups possesses similar nutritional and psychosensorial characteristics, sometimes even symbolic similitude. Adapted to each country, to each type of society, these categories can be used for nutritional planning.

Within the same category each item can replace another. This approach is valid only for the type of society in which the nomenclature has been accepted, a fact which should not be forgotten.

A sociocultural, verbal image of food would suggest a more subtle definition, i.e., brand foods, foods defined by their nutritional properties, short and long-life foods, restaurant foods, prepacked airplane and cafeteria foods, celebration foods, baby foods, geriatric foods, foods for the obese, etc. This nomenclature has yet to be established and must not exclude any of them. Food coming from the earth and industrial food are not competitive but complementary, in the same way that life obtains its equilibrium by alternative factors. Nomenclature should reflect Mother Earth, work, meal-taking habits, and the meaning that man gives to his work and to his bread, in rural areas and in the city. Food language should become the indication of a way of life and should be much more fluid and dynamic than previous nomenclature systems.

Summarizing, we can say that it is possible to define and plan some aspects of food, but in each society there are some powerful irrational factors, which are difficult to seize and which escape all control.

The average food provision is but one of the factors of real consumption by individuals in each micro-society.

Planning can only be valid if it takes into account also nonnutritional factors affecting consumption within each micro-society.

NUTRITIONAL DISEASES

The interpretation of disease in industrial society needs, in our opinion, to be entirely reviewed. The humanitarian slogans of scientific appearance on malnutrition do not correspond to reality. The present conception of disease, born with Laennec, as an anatomical, metabolic, and clinical entity, in which a cause produces an effect, and which a specific drug can cure, represents only part of the reality.

The avitaminosis or specific deficiency which can be corrected by supplementation is, at present, exceptional. For thirty years nutritional diseases have been described and publicized by all sorts of people, but by very few practicing physicians (2, 3).

A drug, a vitamin, or an amino acid does not cure the syndromes of poor eating practices. The anatomical lesions or metabolic problems start a process of interdependent factors which must be considered globally. Furthermore, their causes are numerous and their association is behavioral, bound to psychosocioeconomic life conditions.

At present the most frequent diseases connected with nutrition are weaning diseases, complex malnutrition, plethora-originated obesities, vascular degenerative diseases, alcoholism, and functional pathology.

If we try to determine the really basic factors that we should subdue we find, in every case, that the direct causes are nutrient proportions inadequate for the needs of the individual.

A weaning disease can be cured by a 10 gram daily protein supplement, an excessive weight by a low-calorie diet, a hyperlipidemia by a more complicated diet, and alcoholism by alcohol suppression; but food habits cannot be changed rationally, because they are not rational. To vary calorie or protein intake or to stop ingestion of alcohol we need to refer to motivations and life conditions. Poverty, anxieties of daily life, escapism in food or in alcohol, problems in mother-infant relations, in the work environment, or in the social life, are the determinants on which it is important and effective to act.

It would seem, then, important for nutritional planning to determine the behavioral profile of poverty, of affluence, of family, and of social groups which are connected with the nutritional errors, and to penetrate more deeply into the surroundings in order to find there the motivations and means of changing feeding habits. In this context, disease may be regarded as a kind of "alarm signal" indicating that the society has taken the wrong course.

The present concept of human malnutrition has been elaborated by epidemiologists (not clinicians), by vitamin biochemists, and by economists; but it is not realistic. However, it is used in survey methodology as commended by the World Health Organization. These data, which are far from clinical reality, are useful to planning but not to mankind.

It remains to be determined if poverty and richness are bound together in the social body as the left and right hands are in the physical body and how available food can be distributed among the different strata of society to assure adequate nourishment of the individuals within them.

The fact that disease is connected to the habits of eating means that food signifies health and well-being, or inversely, illness and degradation. Food is, at the same time, our friend and our foe.

The fact that we remain free to choose our food makes it the last refuge of individual freedom. A disease produced by food indicates trouble in ourselves or in our environment.

NUTRITIONAL NEEDS

How can one define the amount of food to correspond with the needs? Here, too, the problem has already been pondered and we shall limit ourselves to listing the points that are essential for planning work (3).

-- Rather than amounts, we should consider margins, which for proteins and calories are ± 50% and for vitamins and minerals, even more; but, if in a homogeneous population some freely choose to eat half as much as others, the compulsion of eating 50% less or more becomes intolerable.

  Each man expresses his personality through the level that he chooses and resents the level being chosen for him. The freedom to choose the amount to eat (within acceptable social limits) remains essential.

-- Above the top margin there is a region producing adverse effects, the toxicity zone. Nutrients, fatty acids, sugars, calories, and alcohol, just as foreign substances, have their danger zones.

-- Below the lower margin, the essential need is not covered and a deficiency may be initiated.

-- The wide spectrum of levels of one nutrient interferes with that of other nutrients and the imbalances result in malnutrition.

-- The needs depend largely upon a series of factors, the most important being the nutritional level during infancy and the seasonal rhythms.

-- From the above it can be easily deduced that if science gives us an insight into such a complicated problem, it does not furnish us with the means to analyze it with precision. The best approach is to observe what man does at a given moment, in a given society, when he is satisfied and judged to be healthy.

To recommend average intakes would be meaningful only if we could affect their distribution in the population.

Man can voluntarily reduce his food intake by half but when he is compelled to do it, he is unhappy. How can we establish intakes from the single to the double, knowing that the individual who could, willingly, adapt himself to either level would suffer when we try to impose it?

This freedom of choice within two nutritional levels at the ratio of 1 to 2 means that man has kept his freedom to a great degree. This individual liberty allows for experimentation that brings out hidden possibilities, inventions, and discoveries which, if valid, will become the models for tomorrow's man, the unforeseen motors of evolution.

It is this freedom which creates the meaning and dignity for each man, which is the basis of real politics. It is freedom which will always define the limits for planning and rationalization, for good and evil, which are eluding science.

CONCLUDING REMARKS

Within the limited scope of this paper, dealing with such a vast subject, it is only possible to give some abstract hints and suggestions.

Planning covers objective food definitions and establishes a cause and effect relationship between food consumption and nutritional diseases. It is oriented to satisfaction of the needs of a population's average individual.

Practice shows that for each of the above points a complementary and dynamic factor intervenes. The objective definition of a nutrient provides only a partial view of all aspects; to it are mixed, in practice, symbolic, emotional, and subjective factors which contribute an emotive note of an entirely different nature.

This emotive note, according to the circumstances, changes the definitions, introducing some factors of indeterminacy and some degrees of liberty which upset predictions. Who could have, three years ago, forecast the natality drop (3-7% p.a.) of the white industrial population and its effect upon baby food consumption? Nutritional diseases are related to nutrient excess, scarcity, or imbalance, but man's behavior is not rational and, therefore, is difficult to forecast.

We should concentrate our efforts on poverty, on opulence, and on social and economic pressures. History shows that man and society seem to find their equilibrium through crisis and disease or, rather, that rational behavior is dictated to man through the experience of his errors. He must feel those errors in his flesh before he can accept them as being reasonable.

Nutritional needs offer us a fascinating example of social behavior. A living society needs to establish the margins of maximum acceptable disorder within its body, as we need to limit our ingesta. These margins are defined by an average level which the planner tries to discover; within them each individual or micro-society enjoys an almost total freedom. The chosen level is obtained by aggregating an ensemble of endogenous factors which must be harmonized individually. The behavioral factors are so numerous and interdependent that each individual possesses a degree of freedom which enables him to try to escape group comportment, becoming perhaps, in the future, the leader of a society which will change its "average individual."

As a consequence, planning is meaningful only as a complement of individual freedom; the latter represents the dynamic; the former, the stabilizing influence in the system.

On the other hand, rigid planning cannot survive because it is only a rationalization of reality and it lacks what we shall call the "policy of individuals and real micro-societies."

Integrating sociocultural factors into nutritional planning should overcome this position and should produce a useful form of accounting to the service of social dynamics. Man and his society would then be devoted to their quest for truth.

The present food situation obliges men to know what is the true meaning of the scientific knowledge, to understand that the objective and logical knowledge, which is science, is meaningless if it does not see its own limitations and if it ignores the understanding of emotions and intuitive, symbolic knowledge.

It is very fascinating that when we look at the simplest act of our daily lives - eating bread together - it is symbolic of man's meaning of himself in the universe.

REFERENCES

1. Trémolières J: A proposed scheme of food behavior. Bibl Nutritio et Dieta 17:144-153, 1972. (Somogy JC, Fidanza F (eds): Nutrition and the Nervous System.)

2. ____________: Nutrition and underdevelopment. In Scheedon M (ed): Progress in Human Nutrition. Westport, Connecticut, the Avi Publishing Company, 1971, pp. 1-28.

3. ____________: Nutrition in public health. World Rev Nutr Diétét 18:275-319, 1973.

BENEFIT-RISK DECISION MAKING AND FOOD SAFETY

William J. Darby, M. D.

President, Nutrition Foundation, Inc.
489 Fifth Avenue
New York, New York 10017

INTRODUCTION

It was the great 17th century philosopher, Francis Bacon, who wrote that

> . . . the real and legitimate goal of the sciences is the endowment of human life with new inventions and riches . . . not to make imperfect man perfect but to make imperfect man comfortable, happy, and healthy.

One may paraphrase this goal as "to provide man with the necessities, facilities and pleasures that contribute to the quality of life." A primary necessity is food. To meet world requirements for this necessity demands the fullest application of scientific knowledge and technologic skills of both agriculture and industry for the efficient production and distribution of food, combined with the application of the modern knowledge of nutritional science.

Increasingly, societies seek to free themselves from the drudgeries and uncertainties of securing this basic necessity for survival through subsistence agriculture, and from the accompanying monotony of life and disease. Such freedom may be achieved by the application of those new inventions of science that endow life with increased riches and thereby enhance the quality of life. "Quality of life" embraces many values: health, education, comfort, aesthetic enjoyment, recreation, leisure . . . and more. Too often persons concerned for one of these values overlook the importance of other values in advocating one or another action. The concept of quality of life is charmingly illustrated by the delightful Spanish-American toast "Salud, dinero amor y tiempo para gozarlo." ("Health, wealth, love, and time to enjoy it.")

THE NEED FOR TECHNOLOGY

May I briefly quote from Dr. George Harrar's Atwater Lecture (7) of 1974 as to the lack of balance between supplies and needs and the potential for fulfilling needs:

> Unfortunately, the world is substantially out of balance with respect to the production and distribution of needed food supplies. It has long been recognized that there are many agrarian nations whose production levels are low and who regularly fail to satisfy their basic food requirements. Most successful agriculture is practiced in temperate climates, where soils are fertile, rainfall adequate and growing conditions, in general, favorable. Where the climate is harsh, water scarce, soils infertile and economic resources limited, agriculture tends to become a subsistence phenomenon. Where there are complications caused by periodic floods, droughts or other natural disasters, the situation worsens. The Third World has long experienced the effects of an imperfect system of agricultural production and a long-term lag in economic development. Where this situation is compounded by explosive population growth, standards of living and the quality of life are woefully inadequate . . . .
>
> Presently, we are faced with new dilemmas, in combination with old ones. The agricultural industry is under increasing pressure to feed our present population plus those (200,000) who join the society every day. In a situation of energy crisis the problem becomes increasingly difficult and complicated, since agriculture

basically depends to a very considerable degree on available energy. Adequate fertilizer, machinery, water systems and electrical power sources in concert make it possible to increase efficiency of agriculture with resultant production benefits. If one or more of those elements is subtracted from the system, reduction in the total food supply is an inevitable accompaniment. The lack of long-range and forward planning, the inability of nations to act in concert for the common good, instability of governments and the totally inadequate emphasis on food production has brought us to our present crisis now being exacerbated by energy constraints . . . .

Fortunately, there is an enormous body of available information, a vast technology, improved biological materials, and an ever-growing cadre of qualified individuals who could give meaning and leadership to any national agricultural development plan. If each nation participates in a global effort to bring about the maximum efficient utilization of its agricultural and human resources for the production of food and agricultural commodities, then it becomes possible to greatly increase total figures worldwide and to plan a system in which food production and distribution can be in harmony with human needs without the convulsions of crisis . . . .

We can have a Green Revolution worldwide. Many of those who have written on this subject seemed to have failed to understand its true origins, meaning, and implication for the future. In fact, the Green Revolution, where it has been applied, has been a dramatic demonstration of the potential of combining all the elements of an efficient agricultural production system and translating them into greatly increased production figures. It has clearly demonstrated the fallacy of some of the earlier claims that many countries are doomed to hunger, disadvantage, and misery on a continuing basis because they are incapable of

> improving their most fundamental requirement, that of an adequate food supply and a proper human diet. The success of the Green Revolution could be repeated again throughout much of the world where both the need and the opportunity exist to apply its principles and practices.

It is reasonable to conclude that no country has maximally utilized all of the potential applications to food of the new technology. But it _is_ apparent that the abundance of food and freedom from want are greatest in those countries where scientific technology has been most intensively developed and applied.

THE BENEFITS OF TECHNOLOGY

The impacts of the application of science are reflected in measurable indices of health. For example, in the United States the industrialization of food production at all levels - with concomitant awareness of nutritional needs and safety - has been accompanied by a virtual disappearance within this century of the classical deficiency diseases: pellagra, scurvy, rickets, endemic goiter, protein deficiency. Infant mortality, a widely employed index of nutritional health of a nation, reached an all-time low in the United States in 1974. By contrast, in many regions where food production and distribution systems are primitive, famine recurrently strikes and high death and morbidity rates continue because of pellagra, iodine-deficiency goiter, protein-calorie malnutrition, folic acid or iron deficiency anemia, and vitamin A deficiency with resultant blindness. These exist in concert with severe and deadly food-borne infections - diarrheal diseases of the young child, enteric infestations, febrile illnesses, and acute intoxications. Half of the children born fail to survive until five years of age. All of these are indices of poor food and water sanitation and contributors to undernutrition and its sequelae.

In a scholarly examination of nutrition and feeding practices and past trends in mortality in infancy and early childhood in European countries from 1841 to 1961,

W. R. Aykroyd, (1) the first Director of the Nutrition Division of The Food and Agriculture Organization of the United Nations (FAO), considered how that experience might guide us in measures for solution of problems of protein-calorie malnutrition today. He quoted a significant passage from a report written in 1861 by Sir John Simon, a great pioneer in public health:

> Factory women soon return to labour after their confinement. The longest time mentioned as the average period of the absence from work in consequence of child-bearing was five or six weeks; many women among the highest class of operatives in Birmingham acknowledged to having generally returned to work as early as eight to ten days after confinement. The mother's health suffers in consequence of this early return to labour . . . and the influence on the health and mortality of children is most baneful . . . . Mothers employed in factories are, save during the dinner hours, absent from home all day long, and the care of their infants during their absence is entrusted either to young children, to hired girls, sometimes not more than eight or ten years of age, or perhaps more commonly to elderly women, who eke out a livelihood by taking infants to nurse . . . . "Pap, made of bread and water, and sweetened with sugar or treacle, is the sort of nourishment usually given during the mother's absence, even to infants of a very tender age . . . . Illness is the natural consequence of this unnatural mode of feeding infants . . . . Children who are healthy at birth rapidly dwindle under this system of mismanagement, fall into bad health, and become uneasy, restless and fractious.
>
> Abundant proof of the large mortality among the children of female factory workers was obtained.

In his epoch-making book, Infant Mortality, published in 1906, George Newman (8) emphasized the extreme importance of suitable infant feeding. "It is not everything," he wrote, "but a greater factor than any other thing . . . . Even the domestic and social conditions are reducible to terms of nourishment." His study revealed that at that time "hand-fed" infants had a lower chance of survival than those who were breast fed with a comparative mortality of three to one. "The excess affected particularly deaths from diarrhea." And it is not surprising that the breast fed infants had a lower mortality since infants deprived of breast milk were given as principal articles of diet: diluted fresh (unboiled) milk; condensed milk, especially skimmed sweetened, diluted, condensed milk; "starch" foods such as biscuits, arrowroot, or bread; and certain poorly designed nutritionally very inadequate "patent" foods. The well-to-do substituted wet-nursing, which also presented problems of limited supply, transmission of diseases from nurse to child, social problems arising from neglect of the wet-nurse's infant, etc.

Aykroyd underscored the obvious similarities of these earlier problems in Europe and those existing today in developing countries.

He stated:

> Whereas there may be a few discrepancies, it is reasonable to suppose that something like the complex of malnutrition and infection that we now call protein-calorie malnutrition was prevalent in the affluent countries until recently, with disastrous effects on infant and child health. Clinical observations and vital statistics show that it has almost entirely disappeared from these countries. Its disappearance has been hastened by the establishment and development of maternal and child health services and centers (which began between 1900 and 1910), better housing and sanitation and higher levels of education accomplished by a general rise in living standards.
>
> The most important factor has been improvement in infant and child feeding, associated with the introduction of safe milk, processed infant foods, mixtures based on cow's milk, and the education of mothers in hygienic feeding methods. Greater reliance on breast milk has played no part in reducing infant mortality in the affluent countries, though it was strongly advocated by

Newman and other authorities. In fact breast feeding has declined almost to the vanishing point during the last 40 to 50 years in most of these countries, a period that has seen a transformation in infant and child health.

Experience thus suggests that PCM can be eliminated in a few decades by the establishment of adequate maternal and child health services, rising standards of living, and hygienic artificial feeding. Examples of this can be found not only in highly developed countries in the temperate zone, but also in poor countries in the tropics, e.g., Barbados, and Puerto Rico. In unusual circumstances, the change can actually take place "at one bound," so to speak. With M.A. Hossain, the author showed that when a community from an underdeveloped country was transferred to a well-developed country and adopted the infant feeding practices of the latter, its infant mortality rate immediately fell to something near the level prevailing in the new country. In this instance the fall took place in a Pakistani community that migrated to Bradford in Yorkshire, and affected the first infants born to women who had left home to join their husbands. The infantile mortality rate in the part of Pakistan from which the families came was about 150. In Bradford it was 45. The new environment meant, among other things, an abundance of cheap cow's milk in various forms and an absence of serious intestinal infections. Perhaps the most remarkable thing was the immediate and almost complete abandonment of breast feeding by the immigrant mothers, in spite of the fact that in Pakistan breast feeding is the accepted and traditional infant feeding practice, and an infant who is not breast-fed has little chance of survival.

He concludes:

Experience in the affluent countries has shown that this complex can be rapidly eliminated by

> efficient health services, rising standards of living, and hygienic artificial feeding. Greater reliance on breast feeding has played no part in its disappearance. The most promising method of attacking PCM in the developing countries is to promote the production and use of cheap feeding mixtures, based on plant foods that fulfill the infant's needs for calories, protein, and other nutrients (1).

Such is the benefit of combining in application our knowledge of the science of nutrition and medicine with that of food science and agriculture.

Other health changes, the so-called diseases of the affluent society, may accompany industrialization; but these are not primarily due to alteration in food patterns. They reflect the profound difference in life styles between the rural nonindustrialized society and the urbanized industrial complex and in part the increased survival time and longevity that results from scientific development. None of these diseases, however, is a stranger to societies that are scientifically untouched. Modern application of scientific knowledge of the health and food sciences and proper recognition of nutritional needs can also reduce the toll of these diseases.

During the earlier development of industrialized food systems some deficiency states arose, but these do not occur in modern times because of the awareness of nutrition requirements and the nutritional quality of foods. The experience in perfecting scientifically-planned artificial infant feeding in the United States may be cited. During the later decades of the 19th century and early decades of the present one, protein deficiency, scurvy, rickets, and iron deficiency anemia were common among infants and young children because of the inadequacies of feeding regimens. The scientific design of the composition of infant formulas with attention to our "newer knowledge of nutrition," combined with hygienic measures in the preparation of food and water and the wide availability of suitable foods for babies eliminated these diseases as scourges. These developments have been accompanied by the remarkable reduction in morbidity and mortality rates among infants and the under five year old group.

The utilization of new technology often conserves nutrient content. It is worth noting that the proper use of chemicals in food production and processing has not resulted in chronic injury to the consumer - an anxiety expressed by many. It permits the most efficient utilization of the potential resources of both agriculture and industry for the most economical production and distribution of nutritious foods. Storage losses are minimized and availability enhanced. Dr. Leif Hambraeus, of the Swedish Nutrition Foundation, and I recently have proposed nutritional guidelines (5) for the beneficial utilization of industrially produced materials.

The economic impact of technologic development is of great concern to governmental planners. One index of the economic impact is the expenditure for food expressed as percent of income. In a cash society this index reflects the relatively greater riches referred to by Francis Bacon - riches available for enhancing the quality of life. In highly industrialized countries of North America and parts of Western Europe 16 to 22 percent of income is expended on food. At the same time nutritional deficiencies are notably milder and less prevalent than in areas with less industrialized production of food where 44 to 80 percent of income is spent for this commodity. Not only is food relatively cheaper in industrially developed countries but often it is cheaper in absolute cost. Application of science often results in a much more abundantly available commodity at a cheaper absolute price, as exemplified by decreased costs of turkey and chicken in the United States within the last two decades.

Preservation and distribution methods erase limited seasonal availability of perishable foods and remove them from classification as "seasonal delicacies." Quality standardization minimizes variation in taste. The resultant psychologic and sensory reactions may reduce the special appeal of a previously rare or exotic food when it becomes commonplace, a reaction that is sometimes misinterpreted by the consumer as an alteration in nutritive value. Conversely, nutritional values may be better preserved by such processing than they are in unprocessed or home-processed foods.

Often there is failure to recognize the enhancement of the quality of life through improved health and economic savings that accrue from technology, but such economic advantages of the future may differ from those of the past. They will be determined in part by the priorities and standards that society sets for environmental considerations - e.g., for waste disposal, for elimination of useful agents that may have some environmental impact, or for eliminating all uncertainties or risks.

THE RISKS AND ALTERNATIVES

Advances in science and technology are never equal on all fronts. For example, modern analytical tools permit the detection of traces of substances, previously unrecognized but in fact long present in our environment and in our food. Current technology does not permit us to define precisely whether there is a meaningful chronic risk associated with this exposure level, but if a minimal risk exists it has long been endured in our ignorance of it.

How much is society willing to pay or to sacrifice in order to remove this uncertainty? Is society willing to forego all food that contains detectable traces of the substance that may be harmless? The *Report of the Secretary's Commission on Pesticides and Their Relationship to Environmental Health* concluded that if one were to attempt to avoid all traces of DDT that can be detected analytically it would mean removal of all animal products from the diet (10). Man has lived in amicable ignorance for generations with certain of these substances. For example, fish preserved from periods long before man added mercury to the environment contained levels of this heavy metal comparable to those found today.

Is the risk of divesting ourselves of a useful agent, such as a needed pesticide, greater than continuation of its use or vice versa? Which decision is more acceptable to society?

Have we more or less knowledge of the risks, including experience in man, of some new alternative than we have of a useful familiar tool? For example, is the alternative the use of an agent like an organophosphate, acutely toxic to an applier, in lieu of one like DDT that has a long record of safety to the applicator?

What are the alternatives? . . . use of another agent?

Is there a suitable replacement?

Is an alternative the increased risk of a deadly microbial toxin like botulism? Nitrates are invaluable in meat processing as a means of reducing the danger of botulism - a decided hazard. On the other hand, there are those scientists who are apprehensive lest nitrates, when misused, might under certain conditions give rise to a carcinogen. While the latter is theoretically possible, there are no human data that point to a carcinogenic action in man. When effective anti-botulism levels of nitrates are employed, studies have revealed the absence of the postulated formation of carcinogens. But we do have known experience concerning the lethality of botulism! The decision, therefore, lies between which of these hazards society chooses . . . a postulated or a known hazard. What degree of benefit will one trade for safety from an uncertain hazard?

Is the alternative an increased cost of production of an important food, such as is the case relative to DES in animal production? . . . or is the alternative a decreased supply of a valuable food?

How certain are we that there is a meaningful risk involved? One of my favorite couplets relative to such uncertainty is that of Hilaire Belloc, who noted in his poem concerning the microbe that although no one had ever seen a microbe, "Oh, let us never never doubt what nobody is sure about!"

Many are the considerations necessary properly to determine the benefit-risk equation (3). Basic to such determination is the recognition that there is no absolute safety. Safety is relative and is in fact only "acceptable minimal risk."

The level of that acceptable minimal risk is a determination that society makes based upon value judgments that seldom if ever are founded on science and strict logic. Value judgments are composites of attitudes determined by history, by cultural experiences, by religious and ethical influences, by economic forces, by experience and by need (4). Indeed, Rousseau in Emile states, "All our wisdom consists in servile prejudices, all of our customs are but servitude, worry and constraint."

Our wisdom varies with social grouping and time. It determines the personal satisfaction of the individual and conditions his way of life, his assessment of "quality of life." The conscious entrance of scientific considerations into value judgments is a very recent phenomenon: its logical employment is unfortunately still a rare one.

It is obvious that there exist professional responsibilities of chemists, physicians, food scientists, agriculturalists, economists, industrialists, consumer spokesmen politicians, statesmen, lawyers, and others involved in formulation of policy bearing on these matters.

National leaders, both political and industrial, must espouse sound and truthful positions on issues of such importance, and these positions must have the benefit of the best scientific judgment available. Participating scientists must maintain objectivity not only concerning scientific evidence, but also relative to their own limited competence to make dogmatic pronouncements on extraordinarily complex societal decisions.

TYPES OF BENEFIT-RISK

These societal policy considerations cannot be decided by scientists alone (2, 3). Benefit:Cost and Benefit:Risk as conceived by the consumer and the producer and the politician are not necessarily the same as conceived by the scientist.

What is Benefit? Benefit is anything that contributes to an improvement in condition.

What is Risk? Risk is the chance of injury, damage, or loss.

What is Benefit:Cost? Benefit:Cost is all the benefits and costs of proposed action . . . a much broader concept than traditional cost:

benefit analysis which includes only economic considerations.

What is Benefit:Risk? Benefit:Risk is that category of benefit:cost in which risk to life and health are important components of cost.

In the analysis of Benefit:Cost or Benefit:Risk, in technology assessment of all sorts, it is obviously necessary to deal with uncertainty (9). But it should be noted that even when uncertainties, in a statistical sense, are resolved, risks to life and health are present in many individual and collective activities.

A consideration in technology assessment is public benefit vs. individual (or private) risk (9). This consideration is inherent in vastly diverse types of decisions. For example, should we continue smallpox vaccinations? Should boys be vaccinated against chickenpox to protect their mothers from infection during pregnancy? Should iodinization of salt be mandatory? Should a man's home be destroyed to make way for a new, safer highway?

Societal activities that result in risks or benefits are of two categories - those in which the individual participates on a voluntary basis and those in which the participation is involuntary, i.e., imposed by the society in which the individual lives, imposed by law, regulation, social custom, religion, etc.

Chaucey Starr, from a broad analysis of Benefit: Costs in socio-technological systems, concludes that:

-- an upper guide in determining the acceptability of risk is the rate of death from disease;

-- a base guide for risk tends to be set by natural disasters;

-- societal acceptance of risk increases in a non-linear manner with the benefits to be derived from an activity;

-- public acceptance of voluntary risks is some 1,000 times greater than that of involuntary risks;

-- societal policy for the acceptance of public risks associated with socio-technical systems should be determined by the trade-off between social benefits and personal risk (9).

Reflection leads to recognition that benefits and risks, both public and individual, may likewise be either vital or non-vital. Dr. Richard Hall has admirably discussed these concepts as follows:

> "Vital" is, of course, concerned with, or manifesting life, and "necessary" or "essential" to life.
>
> A vital risk, then, is a chance of injury, damage, or loss of life. This includes great pain and incapacitation. A characteristic of vital risks is that one can measure them, but not in dollars . . . . The risk of fatality in commercial air travel is about 0.07 deaths/100 million passenger miles. That involved in private automobile travel is about 4.9 deaths/100 million passenger miles. We can measure very accurately (such) changes . . . but we regard it as . . . immoral to put a dollar value on them. True, juries do it every day, but as an after-the-fact compensation. No one ever arranges in advance to sell an arm, or a life. The risks are nonpecuniary.
>
> A vital benefit, then, is something contributing to an improvement in a condition essential to life. Vital benefits can also be measured, but not in dollars . . . . We can measure the effect of hypotensive drugs, and the reduction in an infection due to antibiotic. These, too, are non-pecuniary. Their price has no relationship to their value to one who needs them. (The non-pecuniary nature of vital factors is not widely appreciated by economists nor many politicians.- W.J.D.)

A non-vital risk is a chance on injury, loss, or damage in a non-vital but possibly important way. Loss of comfort or pleasure, inconvenience, frustration, hurt pride are among the non-vital risks.

A non-vital benefit, similarly, contributes to an improvement in these areas of pleasure, convenience, comfort, or ego satisfaction. By contrast, non-vital risks and benefits are usually pecuniary. We measure the value of avoiding or gaining them in dollars and cents.

Of course, the boundary between vital and non-vital risks - or benefits - is not always clear. But the existence of boundary cases is no bar to recognizing and dealing with them as different areas.

As just mentioned, we do not buy or sell vital risks and benefits for a price which reflects their true value. But we will trade off or accept a vital risk for an even greater vital benefit.

Thus, chloramphenical is still the drug of choice for typhoid fever, even though it carries a substantial risk of aplastic anemia . . . .

Note that . . . society does not leave these hazard-laden decisions to the unrestrained choice of the individual. We insert a professionally qualified decision-maker, the physician, into the process, and we support and limit him with the apparatus of the Health Protection Branch of the FDA.

And even when qualified, there is a self-exclusion. 'The man who is his own doctor has a fool for a patient.' This cuts both ways. It not only excludes the amateur physician, but excludes the qualified physician from serious practice on himself or his family where distortions of judgment can easily occur.

So, acceptance of vital risks to gain a still greater vital benefit is entirely proper, if we

> first provide the legal and moral framework to govern these choices (6).

In the context of benefit-risk decision making relative to safety and adequacy of food we must consider the application of science to food production at the levels of agricultural and animal husbandry, of storage, preservation, distribution and use, of industrial processing and marketing, of quality control, and of regulatory protection and standardization. These benefits are sometimes vital, sometimes non-vital. I need not argue the case further. Proper assessment of these scientific developments must include identification of public and individual benefits and risks, vital and non-vital in nature, voluntary and involuntary, regulated or controlled, and uncontrolled.

Unwise regulatory or legislative constraints that prevent the application of science and technologic knowledge to increasing or improving food production and prevention of spoilage or losses can have disastrous world consequences. Such risks as may result from not having the use (benefits) of an agent should be weighted in decision-making against whatever risks, certain or uncertain, that may be responsibly judged to pertain from laboratory findings in relation to any particular useful agent.

What are the risks of divesting ourselves of the useful agent? Can we afford them? What are the trade-offs?

Can we justify denying the benefits of use because of postulated extrapolation of experimental findings to man? Who is "we"? Scientists?

THE ETHICS OF BENEFIT-RISK DECISION MAKING

What is the nature of the ethical problem facing us in decisions related to food safety? The Citizens' Commission on Science, Law and the Food Supply (2) defines it as follows:

The Central Issue

> To what extent should consumers be exposed to either known or possible added risks of various potentially toxic substances, including

carcinogens, mutagens, and teratogens in their foods? (To this we might add, to what extent should consumers be exposed to shortages, to increased cost, or nutritional deficiencies?-W.J.D.)

Preliminary Statements of an Ethical Guideline

Our guideline should reflect the proper ethics in the face of uncertainty.

When faced with alternatives, we should choose that alternative whose permitted worst outcome is better than the worst outcome of any other alternative, where "worst outcome" is defined as a cumulative net balance of risks and benefits.

The upper limit of the overall risk should be no greater following a change than the risk of natural disease demonstrated to be causally related to existing patterns of food use.

Consumer Risk is not an Isolated Event

Risk in food is one part of a closely-knit set of values fixed in modern man's life-style. Therefore, no useful answer to the question of acceptable risk to man can be given without taking into account the other parts of the whole life-style patterns.

The Citizens' Commission discusses the expression of the ethical problems in government regulation and the ethics of accommodation as follows:

Formation of policy decisions rests upon the assumption that questions of policy are not solely scientific questions but are matters which also must involve society's value judgments.

If these were solely scientific questions, then the scientists should propose the policy.

Part of the dilemma is that scientists disagree among themselves over the meaning of some of the critical data; such disagreements may stem from:

-- the scientists' lack of breadth of outlook,

-- use of an unsuitable or inappropriate animal model,

-- unjustified extrapolation of experimental conditions to real-life situations,

-- honest disagreement over implications of scientific data.

Another part is that the conclusions that can be drawn from all available data are often uncertain.

Under present circumstances, a single experiment by a single laboratory investigator can provide the basis for removing a product from the market. Scientific consensus is needed on how much experimentation is appropriate.

Facts do not necessarily assess themselves; they need to be evaluated and interpreted in terms of some basic values to which our society is committed.

According to the Citizens' Commission, creation of policy requires that some accommodation be achieved between:

-- expectations of those who buy and consume food products,

-- the principal interests of producers,

-- the information of the scientists, and

-- the general interests of society at large.

The Commission suggests that policy decisions should represent a consistent, coherent, and rational ordering of the concerns of the four interest groups (consumers, producers, scientists, and the general public).

In this context, scientific information is of critical importance.

Equally important is a reasoned analysis of the implications of this information, both certainties and uncertainties.

Consumers' expectations should be submitted to the same critical analysis and evaluation. To do this effectively, appropriate methodology should be developed to ascertain what consumer expectations are with regard to major problems.

Undue pressure upon regulatory agencies by whatever interest group must be prevented.

In any society the regulatory mechanism should, to the extent possible, serve the diverse interests of both the poor and the affluent.

In the end, the definition of acceptable risk should be hammered out in terms of clear rational principle, taking into account all the relevant interrelations and having in mind the need for periodic review and revision of regulations as new information and uses emerge.

To achieve these objectives, regulatory agencies should be supported by highly competent, independent advisory boards whose members are drawn from a variety of professions and activities and whose disciplines or careers have prepared them to deal with either the special problems or the societal problems.

As workable ethical guidelines for acceptable risks, the Commission suggests the following:

There must be no substitute for an aggressive reduction of all known vital hazards whenver they become known.

The general overall principle, as set forth in the preliminary statement of an ethical guideline, should be used as the starting point for specific guidelines.

-- Specific guidelines should be tested by the general principle.

-- Specific guidelines should be applied only after available data have been carefully and collectively considered.

The Commission concludes with a detailed statement of the general principle which should govern all benefit-risk decision making in the area of food safety:

> When faced with alternatives, we should choose that alternative whose worst outcome is better than the worst outcome of any other alternative.
>
> Using this formula, the alternative of absolute safety would usually be "worse" than the alternative of using controlled chemical substances with their implied risks properly evaluated and controlled.
>
> To make this general principle more useful, we need more refined information concerning the cost of risks and the value of benefits.
>
> Even after we possess reliable quantitative data regarding the costs of risks and benefits, an ethical principle, as given above, is needed to determine whether and under what circumstances a particular implied risk is justified.
>
> The phrase "worst outcome" in this general principle should be construed as:
>
> -- a risk no greater following a change than the risk of natural disease demonstrated to be causally related to existing patterns of food use.
>
> -- the cumulative net balance of of risks and benefits in the population as a whole, not the

worst outcome in a single individual case.

The decision of the acceptable risk should be made at two levels:

-- legislative: broad general guidelines should be defined by Congress (or appropriate legislative body - W.J.D.)

-- administrative: specific decisions should be made by the appropriate administrative agency with the support of advisory boards as described above and with the assistance of competent surveys of consumer expectations.

As the consideration of ethical principles pertinent to the determination of acceptable risk develops, it may be essential to re-evaluate and further refine the general principle above.

Only within the framework of such broadly conceived deliberations can society balance the benefits and risks of the application of science to agriculture and food production. Failure to do so ultimately can lead to serious impairment of productivity and to degeneration of the nutritional health of people throughout the world.

REFERENCES

1. Aykroyd WR: Nutrition and mortality in infancy and early childhood; past and present relationships. Am J Clin Nutr 24:480-487, 1971.

2. Citizens' Commission on Science, Law and the Food Supply: A Report on Current Ethical Consideration in the Determination of Acceptable Risk with Regard to Food and Food Additives. New York, 1974, p. 28.

3. Darby WJ, Acceptable risk and practical safety: philosophy in the decision-making process. J Am Med Assoc 224:1165-1168, 1973.

4. __________: Nutrition and new food technology. Summary of meeting, Science and Man in the Americas, Mexico City, June, 1973. Am Assoc Adv Sci, June, 1973.

5. __________, Hambraeus L: Proposed Nutritional Guidelines for Utilization of Industrially Produced Nutrients, 1975.

6. Hall RL: A modern three R's - risk, reason and relevancy. J Can Inst Food Sci Tech 6:A17-A21, 1973.

7. Harrar JG: Nutrition and numbers in the third world. Nutr Rev 32:97-104, 1974.

8. Newman G: Infant Mortality, a Social Problem. London, Methuen, Ltd., 1906.

9. Starr C In: Perspectives on Benefit-Risk Decision Making, Benefit Cost Studies in Sociotechnical Systems. Washington, D. C., National Academy of Engineering 1972, pp. 17-42.

10. United States Department of Health, Education and Welfare: Report of the Secretary's Commission on Pesticides and their Relationship to Environmental Health. Washington, D. C., Government Printing Office, 1969.

FOOD, TRADITION, AND PRESTIGE

Igor L. de Garine, Ph.D.

Maitre de Recherches
Centre National de la Recherche Scientifique
Maison "Pargade", 64 Lasseube
Pyrenees Atlantiques
France

Since most studies of prestige foods deal with urban groups, we shall focus our attention on traditional rural societies. Among the various symbolic values likely to affect food behavior, prestige is frequently mentioned. In most human societies, the fact that one eats or offers such and such a food is associated with status considerations, the result being that certain foods and dishes are more prized than others.

Man is perhaps not the only animal capable of using nutrition as a means of differentiating between individuals and groups. One could imagine, for example, that the rivalry which opposes carnivorous animals after the same prey may have the same result. Man is, however, undoubtedly the only animal capable of eating food whose organoleptic properties are mediocre simply in order to display his economic prosperity to his fellows. He is also the only animal which, in order to exhibit his social position, will deliberately forego food which is wholesome but associated with low social status.

Some dishes are meant to be eaten within the family circle; others are openly offered to guests. Some foods reflect financial stress, others symbolize affluence. Man can eat more than he requires to satisfy his hunger, simply to impress other people. The importance of prestige food is not a question of how much pleasure is derived from its consumption, but of how much social recognition it confers. Perhaps the gourmets of Ancient Rome did experience some particular gastronomical pleasure when they feasted on larks' tongues, but the anecdote about Cleopatra's pearl was certainly not concerned with its gastronomical aspects. If one believes a recent article of R. Oliver (1) it was ". . . at the time, considered to be good taste to give guests the sum spent on the banquet to which they were invited. When the cost of the meal was announced, one of the guests expressed his doubts regarding the enormous sum. Cleopatra then took the exquisite pearl from around her neck, threw it into the dish, and said, 'Does this make up the amount?'"

The terms "prestige food" and "food prestige" may, in fact, be interpreted in many different ways.

FOOD FOR DISPLAY

This may be food eaten by individuals and groups on particular occasions in order to exhibit their status, especially on a socioeconomic basis, and even to improve their social position through the display of their prosperity.

Over fifty years ago, Marcel Mauss, (2) in a famous essay, demonstrated the part played by gifts - and especially food gifts - in the establishment of social ties between individuals and groups. Most collective ceremonies which include two or more social groups are accompanied by the consumption of food which is different, both in quantity and quality, from the ordinary daily fare. Most events in the religious, social, and political life of a community, and most of those in the life cycle of individuals belonging to a family group, have food connotations. Collective ceremonies uniting two different social groups are generally of a competitive character: the hosts display their prosperity, putting their guests under an obligation to return the invitation on at least the same level unless they are prepared to admit their inferiority. Such challenges can be observed to a greater or lesser extent in all societies. In some areas they have polarized economic and social life and led to violent economic contests which alone can confer a higher status on individuals or groups. Contests of this kind, which took on an almost caricatural form, were studied at the beginning of the century, among the Indian tribes of northwest America, where they are known as "potlatch." In these societies, which were mostly freed from the constraints of subsistence economy, the chiefs enhanced their prestige by issuing to one another invitations during which considerable quantities of valued goods - blankets in particular - were distributed or publicly destroyed, and large amounts of food were eaten and given away. The guest then had to return the invitation, when a greater number of riches would change hands or be destroyed, or else publicly acknowledge his inferiority (3).

To the despair of modern planners, prestige economy and the conspicuous consumption of food are still very important in most developing countries and, in many cases, savings which have been scraped together over years are swallowed up in a moment. In 1961, among the rural population in the Deccan in India prestige expenditures incurred in family celebrations were as follows: birth - 9% of the annual income; hair shaving - 12%; circumcision - 17%; first menstruation - 23%; weddings - 192%; funerals - 58% (4). In the small town of Thiès, Senegal the cost of a Muslim naming ceremony, most of which is spent on meals and delicacies to be offered, varies between 5 and 10% of the annual income (5, 6).

In Chad, among most of the animist groups of the South, the sum spent on buying cattle which will be eaten ritually at the funeral of a close relative grossly exceeds the annual financial income.

The occurrence of such ritual meals is not left to individual initiative; it is dictated by events and the expenses borne by communities. Such meals have, to a certain extent, been perpetuated to the present day in connection with christenings, first communions, weddings, and funerals, which constitute today the only occasions when groups beyond the nuclear family gather together.

It is in the same secretly competitive perspective that one should consider the network of invitations to lunch or dinner which, until recently, made up the essential of social life in the western world and which were, in fact, dissimulated economic contests no longer carried out on group initiative but by individuals following carefully set down rules. In this respect, 18th century writers in France are unambiguous. Grimod de la Reynière wrote: "A man with a position who wishes to establish a clientele; a fine poet in search of people to praise him; the ambitious man who needs protectors; the rich man who wishes to be talked about; and the minister who wants to be considered a great statesman, have no better way of achieving their aim than by inviting people to a meal. The table is the center around which all reputations are formed . . . ." (7)

The same author recommends that guests express their satisfaction in terms of social recognition. ". . . Must one not show some gratitude towards him who goes to so much trouble to feed us his wealth. Instead of bantering and outraging him, let us pay our share in happy speeches, fine words, erotic verses, witty repartee, amusing and short anecdotes." (7)

Such recognition is exactly what the Indian chief of British Columbia had in mind when during a "potlatch", he overwhelmed his guest with bundles of dried salmon before an audience who witnessed his generosity. He was, however, relying more on the abundance of his gifts than on their flavor. In societies emancipated from subsistence economy and where a clearer social hierarchy operates, attention is paid to quality rather than to quantity, although after reading a 19th century menu, one may wonder about the sensation of repletion which must have accompanied such gastronomic meals.

PRESTIGE AND GASTRONOMY

The classics in gastronomical literature of the 19th century in Europe are quite clear on this point: commoners devour, the nouveau-riche are gluttons, only gentlemen - even those who are short of money - may be connoisseurs. The satisfaction of the primary need of hunger has become the art of the table. The envied title of "gourmet" is reserved for an elite who, no doubt, possess certain natural capacities but who have, above all, accepted undergoing a long initiation and derive their pleasure on an intellectual rather than a material basis. Being rich certainly does no harm but it is not essential.

"Money alone is not sufficient to ensure a good table. Everything depends on the care, knowledge and studies one has made of all the aspects of the art of food. The rôle of Amphitryon necessitates, like the others, an apprenticeship . . . and in order to fill it well, one must add to a good education as deep a knowledge of men as that of good food . . . . Moral qualities are no less necessary . . . ." (7)

The greatest care is taken to situate on a philosophical level what is simply a food hedonism which one refuses to admit. Few eras more than the French 18th and 19th centuries ever connected so intimately food consumption and social position, the latter of which was founded on different criteria. Financial riches were important, as were knowledge and culture, which until then had been the prerogative of those social classes to which one was admitted almost exclusively by birth. Brillat-Savarin invented a series of tests which he named his "gastronomical test tubes," and which were graduated according to the financial income of the upper social strata. " . . . The strength of the tests should take into account the faculties and habits of the various social classes . . . ." (8) Within each class, the presentation of a set gastronomical meal should provoke from the subject such signs of appreciation as to enable his classification as a guest worthy of an invitation or not. In other words, one does not feed pearls to pigs and certainly not roast woodcock to a swineherd. A small investor does not react in the same way as a financier or a minister.

During the 19th century, the development of gastronomy as an element of social prestige appeared as an attempt to transfer the pleasure derived from the feeling of repletion and the satisfaction of primary hunger needs, to a purely qualitative appreciation of the organoleptic properties of food and the variety of flavors. In social terms, it can be interpreted through the rationalizations it brought about, like the rearguard combat carried out by aristocracy against a middle class rendered more and more prosperous by the beginning of industrialization. Their money was willingly accepted but, in order not to appear ridiculous, the middle classes had to abide by a code laid down by gentlemen - mostly blue-blooded. Brillat-Savarin even used the word battles to describe the arguments which took place between the financiers and nobles in the closed circle of Parisian gastronomy (8). This aristocratic concept underlies one of the most frequent interpretations of the relation between food and prestige.

FOOD, PRESTIGE AND ARISTOCRACY

It is in this perspective that D. B. Jelliffe's definition, published at the Seventh Nutrition Congress, should be considered: ". . . All cultures have prestige foods which are mainly reserved for important occasions or, even more, for the illustrious of the community . . . . They are usually protein, frequently of animal origin. They are usually difficult to obtain, so that they are expensive and relatively rare . . . . Lastly, and of much significance, they may quite often have been long associated with the dominant socio-historical group . . . ." (9) The application of this definition to developing countries would seem a little dangerous as it gives the impression that there is in each of these countries an aristocracy which serves as a model for the rest of the population and whose prestigious style of life, and in particular its food, is envied. This may lead to the over-simplified conclusion that, once this model has been defined, it will be easy to take the appropriate measures to improve nutrition. Such a proposition cannot be applied with equal success to both rural and urban societies.

Rural Areas

In rural areas, since the economy is only partially monetarized, the traditional models continue, in the main, to operate. Few societies possess the medieval style of hierarchy. The differences between social categories, when they are present, are usually based on birth and are impossible to clear except by accession to socioeconomic power. The various classes and castes are interlocked in a system of gifts and counter gifts which accentuate their complementarity and where each individual is surrounded by prohibitions and taboos: each person has his own specialty and his own food. However, an abundance of choice food is not necessarily associated with the upper social strata. In India the Brahmin can hardly be considered to have a gourmet's attitude toward food. In Senegal the Muslim priests are those who have the lowest consumption of protein foods (10). If one can use the word prestige in this context, it is in relation to their modest attitude regarding food and it will be transmitted as such to other categories. Few traditional societies have reached the degree of social differentiation at which an aristocracy emerges and is characterized by a particular range of food. There are, of course, differences according to economic status: a village headman eats more abundantly and probably consumes more proteins than his poorest villager, but he eats very much the same food. In a subsistence economy system, they are both submitted to the same climatic hazards. In many rural societies, similitudes and ethnic solidarity with regard to food supersede the dietary differentiations which may result from the establishment of a hierarchy.

A traditional society may be defined by the feeling of belongingness experienced by its members through having the same staple food and food prohibitions. In Guinea, the Kissi are nicknamed "the rice people." The Massa of Cameroun are called the "red sorghum eaters," the Moussey, "those who avoid horsemeat."

Cultural authenticity is the first stage at which prestige may be observed. The status of human being properly inserted in the framework of a society is initially conferred by the consumption of the staple food - the daily bread. "Tell me what you eat, I'll tell you what you are." Brillat-Savarin (8) could not have guessed that his aphorism would be applicable so deeply and in such exotic places. It is by consuming the group's staple food that its members display their belongingness to the society and their consubstantiality with Mother Earth, so much so that R. and L. Makarius (11) spoke of exogamy based on residence and common partaking of food. Among the Massa of Northern Cameroun, an immigrant who is unable to eat the produce of the land of his hosts without undergoing a ritual, will find himself, once the ritual is accomplished, submitted to the same exogamic taboos as his hosts, and consequently will have to find a wife outside the group.

The envied, prestigious diet is not that eaten by an elite which is as yet ill differentiated. It is, above all, the daily diet which was consumed by the ancestors and is based on the staple food, granted by the supernatural powers, which confers man's condition within a given society. The members of a traditional group do not notice the monotony of their diet. They believe that they eat the food which is best adjusted to their specific nature. One must, therefore, include what Jelliffe (9) calls "cultural superfoods" as one of the most important prestige foods having a bearing on the evolution of the food behavior of those who eat them. We should remember that we are dealing here with the most frequently eaten staple foods, those which enter into most of the rituals and are associated with the religious and philosophical concepts of their utilizers. Inversely, food xenophobia is a characteristic of societies which hesitate to qualify their neighbors as humans because of the food they are willing to eat. The French call the Italians "spaghetti eaters," while they themselves are considered as repulsive frog-eaters. The spirulina, a protein-rich micro alga discovered in Chad, which roused nutritionists' interest a few years ago, is the food of the Kanembou tribe. However, since these people are regarded as the slaves of most of the populations of the area they occupy, it is probable that if a program were launched to promote the product in Chad, it would run up against ostracism.

The oppositions which lead to certain food or diets bearing a status connotation do not occur only between social categories within a given society but also between whole societies which are distinct, enjoy different cultures, and whose diets may present deep structural differences.

The military, technical, and political pre-eminence of the Fulani over the paleonegritic populations of Northern Cameroun have led the latter to adopt a large number of Fulani cultural features, including their staple food. The history of developing countries is little known; it is also rich in confrontations which have led to subordinations which have rarely been openly admitted but may still have daily repercussions regarding food. In the traditional state of Burundi, dominated by the Tutsi and Hima shepherds, the Hutu and Twa farmers and hunters were undoubtedly nostalgic for cattle and its by-products. As H. G. Bartlett writes: "History reveals that the members of one ethnic or sub-ethnic group frequently become paragons for another." (12) In rural, non-monetarized, traditional circles, a glance at the regional history nearly always enables one to detect which groups played the main role in cultural change.

Urban Societies

In those urban societies where a fraction of the population recieves a regular salary, food becomes a permanent social status factor. This differs from that which can be observed in traditional rural environment, where it suffices to prove on a few specific occasions one's prosperity in order to assert one's position. The traditional food pattern is modified and new foodstuffs and dishes appear - often easy to prepare and symbolic of a certain modernism. It is, however, probably possible to improve on assumptions based on the opposition between modern and traditional food, cheap and expensive food, imported and home-produced food, "proper" food and bush food, protein-rich food as opposed to carbohydrates, and scientific food as against natural, wholesome food. In Niger rice has become the staple diet of the privileged urban classes. Town dwellers display their attachment to their African authenticity by eating, on Sundays, the traditional dish of slightly fermented millet and do not hesitate to bring in from the country an elderly relative to prepare it. Here, we apparently observe a return to the past, a desire for natural products symbolizing the Golden Age when everything was provided by Mother Nature.

Monetary income becomes an important factor in the stratification of groups and one witnesses horizontal cleavages into social classes whose ladder can be climbed, thanks to economic riches. This is where Bartlett's proposition fits in: "The behaviour and possessions of upper classes and castes within a society commonly become the focus of goal striving by their inferiors." (12) It is evident that in most African towns a common food pattern is in the process of elaboration and that certain foods and dishes are gaining recognition simply because they are eaten by the upper socioeconomic categories, who are the trend setters. It would, however, be a mistake to imagine that there is perfect homogeneity within these categories - the average consumer, be he Camerounese, Indian, or Brazilian remains a mere abstraction. In Cameroun, the prestigious menu of a high-ranking official Eton from Yaounde, comprising a large proportion of tubers and steeped cassava, would appear as a pure abomination to a Fulani deputy from the North. They would probably reach an agreement by consuming a European-type meal, which would please the marketing specialist but should also incite him to be prudent. Unlike what can be witnessed in urbanized western societies, where regional differences have faded away, the horizontal division of society into socioeconomic classes is recent in most developing countries. It is superimposed upon strong vertical cleavages according to regional or ethnic origins. In spite of government efforts towards national unity and the suppression of tribalism, there is still competition between most of the important ethnic groups who wish to seize political and economic power and impose, more or less consciously, their own cultural patterns. A kind of pecking order is installed among the various entities. The common denominator which finally prevails bears the mark of the society which is enviable for historical reasons or because of its present-day political or economic dominance.

If many of the problems common to large urban centers apply to Dakar, Senegal, and if behavior patterns derived from the industrialized societies hold an increasingly important place there, they are, nevertheless, filtered and modified by the local trend setters, the evolutional Wolof, whose everyday rice and fish have become the national fare. The whole of the acculturation process which affects the populations of Niger and Northern Nigeria is influenced by the cultural style of the Hausa, who have in turn compromised between western models and their own traditional society. It is just a matter of time before the Fulani culture is assimilated by most of the animist groups of Northern Cameroun, which will bring changes in the food patterns and serious modifications of the local food economy. In this respect, one might write that the evolution of a country's food habits is related to its political evolution. The number of trend-setting societies to be found in most of the nations in the process of unification are limited. A detailed study of their traditional food patterns and their changes would permit certain forecasts and avoid the rash extrapolation of the situation found in western industrialized society.

THREE CONCRETE EXAMPLES

The way food accentuates differences in social standing varies according to the accepted scales of values operating in a given society. The knowledge that prestige foods are more often based on animal proteins than carbohydrates does not convey much about food consumption trends. A detailed analysis of a society's activities is necessary in order to determine its scales of values, which are often contradictory.

We propose examining the case of three populations of Central Africa: Moussey, Massa, and Toupouri. They live in close contact and, therefore, justify a comparative approach. This example will also demonstrate the level at which research should be undertaken on the sociocultural aspects of food and nutrition with a general theoretical framework similar to that described by Dr. Simoons in his paper but within a concrete local situation.

The three groups in question are located in Sudanian climate, on both sides of the Logone River, overlapping the Chadic and Camerounese frontiers. They live close to Islamised populations, the most important of which are the Funlani. Each group comprises at least 100,000 souls and would justify in itself a regional development project.

The Moussey

Primarily farmers, they cultivate, on poor soils, their staple food which is early red sorghum (Sorghum caudatum), a little bullrush millet, some eleusine (Eleusina coracana), sesame, and a few pulses. Food gathered in the bush plays an important part in their diet, which also includes small quantities of cassava. Animal husbandry is limited to small livestock: goats, sheep, poultry, and the small ponies which constitute one of the most valued assets of their culture. Fishing is a seasonal activity, whereas hunting used to play an important rôle in the food supply and conferred a certain amount of prestige. The capture of big game was a great event, accompanied by a sacrifice. The hunter was entitled to have a red-painted tree trunk placed on his tomb after his death. Most of the men's mottoes refer to hunting: "My name is Corporal, I am a strong man, I kill game for my father." "My mother, who belongs to such and such a village, sits down and opens wide her thighs to receive my prey." "Your parents eat the shoulders - mine the legs."

The Moussey, who have only one high-yielding crop, are at the limit of subsistence. Each year they undergo a period of restriction and have suffered severe famines. The fact that they have invented a special implement, a pottery oven (Djodjoka) used to hasten the ripening of sorghum which is picked while it is still green, is symptomatic of their precarious position. Their oral literature also reflects this aspect, food and famine being frequent themes. The leading character of these tales is a glutton named Kada who, in his voracity, goes so far as to deprive his children and even devours their fingers when they offer him food. He is, however, always punished for his greediness.

Outside the family group there is little solidarity over food but one should avoid being in a position where one has to refuse food to a hungry person. "We call you but you don't hear." In other words, "When someone calls you in your compound, you pretend not to hear to avoid sharing your food." The months of the rainy season are named according to food shortage. Children are strictly forbidden from wandering from one compound to another for fear that they might beg for food. Greediness and plumpness are not prized. In this egalitarian society the status distinctions show up between the poor and the rich: the former are hungry during the winter, not the latter.

The notion of the quantity ingested is more important than the quality. Eating contests are frequently evoked in the oral literature. For example: One day the trickster, Kada, met a being who resembled a half-man and whose name was Jing-guiling-jinga. Sure of winning against a half-man, Kada challenged him to eat as much game as himself. They went hunting and came across a partridge. "Shall we kill it?" asked Kada. "No, replied Jing-guiling-jinga, "it's too small." Next it was the turn of a hare, then an antelope, a hippopotamus, and finally an elephant. At last Jing-guiling-jinga said, "Yes, that will do."

They killed the elephant and cut it up. Kada ate first, he swallowed a whole leg. Jing-guiling-jinga gulped down the rest of the elephant.

When they came to a pond, Kada drank first and the level of the water went down. His companion drank and drained the pond dry. After a while, Jing-guiling-jinga exclaimed, "Kada, my friend, you should hasten your step; my stomach aches and I have to relieve myself. I shouldn't like you to be bogged down or drowned." Thus Kada lost his wager.

Celebrations. Prestigious food consumption occurs mainly on the collective occasions related to:

-- the religious agricultural calendar: millet harvest, first fruits offerings, New Year's feast. Abundant quantities of red sorghum are eaten, accompanied by meat-based sauce; millet beer flows generously. This feast will show the neighboring populations the favor bestowed by the supernatural powers, who protect the hosts and authorize such a spread.

The religious feasts promote several types of food:

-- red sorghum - the daily bread, and bullrush millet - white and tasty

-- the early crops such as eleusine and groundnuts

-- the meat of domestic animals which are a compulsory part of the sauces; in decreasing order of preference: beef, mutton, goat's meat, and poultry

-- sorghum beer, the confection of which demonstrates the housewife's skill

-- social celebrations: the above-mentioned feasts also provide the pretext for individual invitations which the heads of family issue to guests of their choice in order to display their personal prosperity. All the social events, and particularly those which mark the life cycle of individuals - birth, initiation, marriage, death, - are accompanied by plentiful sacrifices and feasts which establish the social status of the

> group which is entertaining. In order not to lose face, it is essential to appear to be prosperous and lavish with food even if it is only on a fleeting basis.

Funeral feasts are those which give rise to the highest amount of conspicuous consumption. A son must kill a considerable number of cattle for his father's funeral. Sons-in-law also have to bring cattle or some other important gift to the burial of their mother- or father-in-law. The passing away of an eminent person leads to the slaughtering of dozens of head of cattle. These ceremonies are accompanied by feasting, drinking, and a certain licentiousness dictated by tradition.

Parallel to these compulsory celebrations, there is another set of quite optional rituals which contribute to enhancing an individual's social position and where feasting again plays a large part. These are the magico-religious associations: possessed people and initiation groups. In these restricted circles, one may or may not be allowed to join in the ceremonial meal according to one's grade within the group. In this case, the gastronomical aspect is very important since in the societies under consideration, the sacrifices of domestic animals constitute the sole occasion for meat consumption.

The initiation of the young men is particularly relevant. Departed souls are hungry after their death. According to oral tradition, initiation was invented by the ghost of the head of a family who had come back home furtively in order to eat and was surprised by his children. He told them to build an altar in the bush, to begin the ritual and feed him with their offerings. The key of the initiation ceremony lies in the fact that the noninitiated must firmly believe that the food is eaten by the supernatural powers, instead of by full-fledged initiates who make fast work of it behind the scenes. It is only once he has passed the second degree of the initiation rites that a neophyte is allowed to participate in the earnest feasting which takes place in a secluded area of the bush and where meat - preferably greasy - is consumed.

Periodic feasts also mark the ceremonies organized by the chiefs of groups of possessed people. The latter enjoy a high status within the society and are often called upon to display their wealth.

The precautions taken by a group of initiates coming home from a feast, the early-morning offering of the first fruits by the traditional priest, the surreptitious way in which a wife withdraws a titbit from the meal to offer to her husband when he comes to honor her at night, all corroborate the valorization of a furtive type of food consumption which reflects the prevailing insecurity.

As regards gastronomy, the technology hardly allows housewives to distinguish themselves. They succeed in doing so, however, in the preparation of the sauces which accompany the sorghum staple, and which must include meat or, at least, dried fish, and should be greasy, which is not surprising in a society whose diet is very poor in lipids.

Wild products gathered in the bush have their place among favorite foods. For instance, a wild tuber (*Tacca involucrata*), which necessitates a long preparation, is much sought-after.

The Moussey refer to themselves, as opposed to their neighbors, as "the people who do not eat horsemeat," the possession of ponies representing one of the main assets of their social life. Their neighbors, the Massa and Toupouri, consider them to be slaves who eat accordingly, and stigmatize their poorness by saying: "The Moussey eat the mucilage of a tree in the bush (*Grewia mollis*), they haven't even any sorghum gruel to go with it!"

The Moussey provide the example of a common type of traditional society in subsistence economy where the relative equality of resources between the various individuals prevents the elaboration of a scale of values based on food prosperity. They are reduced to enjoying occasional meals with a high protein content and to envying their more prosperous neighbors, the Toupouri, whom they, nevertheless, mock because they eat dogs and the rotten forelegs of cattle, and the Massa, who eat frogs and "even crunch their bones."

The Massa

The Massa enjoy more diversified food resources: they cultivate their staple, early red sorghum, and also engage in profitable fishing in the Logone and its tributaries. But, above all, they are herders, who are cattle-crazy and whose social activities are entirely centered on the possession of cattle and the prestige they confer. The prestigious uses of cattle are the following:

-- bridewealth - ten cows must be paid to the father-in-law in order to obtain a wife

-- cattle-lending (golla) - by lending a cow to somebody, one not only expresses one's confidence in the person who will drink its milk; one also displays one's riches and extends the network of one's social relationships, thus creating a clientèle

-- the gourouna - ritual fattening. This is an original way of resorbing food surpluses and of stressing the economic prestige of individuals within the society. At certain periods of the year, the young men who have not suffered a death in their family withdraw into camps where they drink cows' milk. They lead a very different life from the usual, look after the herds, and consume large quantities of milk and sorghum porridge. They do no work but spend most of their time training for the dances and wrestling contests which mark most of the religious occasions. They also lead, outside their family circle, a very busy love life.

Stoutness is esteemed among the men, as is gluttony. In the camps, milk and millet are wasted to display wealth and even used as cosmetics. The numerous songs composed by the gourouna boast of food, meat and above all, milk, which is considered "preferable to fish, which is smelly."

Although one often comes across the same themes as among the Moussey, the Massa's oral literature is dominated by the gourouna, depicted as strong, plump young men, the heroes of chiefs' daughters, whom they inevitably seduce in the end!

The Massa are more prosperous than the Moussey, although their position is questionable during the rainy season, when they undergo a shortage period, which may be aggravated by the rising of the Logone.

Wealth in terms of food is not translated by a mere opposition between "those who eat their fill, and the others" but in terms of heads of family who own a herd of cattle which they know how to manage and which procures them wives and daughters, as well as milk, as opposed to the others who are unlucky enough to belong to a lineage where mostly boys are born, and are, thus, deprived of cows and their inherent prosperity.

In addition to the gourouna, the Massa have the same religious and social feasts as the Moussey, but have no magico-religious institutions for possessed people. The highly secret initiation which they practice combines numerous themes dealing with food and milk. When a novice is killed during the initiation session, his death is said to be due to his gluttony. "He has been taken off by the female spirit of the initiation. He forgot to let go of one of her ten nipples when she returned to the sky." Here there is no furtiveness. Food, together with quantitative aspects - plumpness, gluttony together with strength, and placidity, are all valued. As regards quality, milk emerges as the most prestigious food of all; the red sorghum which goes with it is a gift of the Gods and is present at all the celebrations.

Toupouri

Among the Toupouri the food situation is similar to that of the Massa. In addition to red sorghum the Toupouri grow several varieties of pricked-out white sorghum (S. guineense and S. durras) which they probably adopted from the neighboring Fulani. They have at their disposal at least three high-yielding cereal crops and actually enjoy an abundance of grain, (13) much of which is wasted in the gourouna institution, and the traditional celebrations for which considerable amounts of beer are brewed.

Here we are no longer faced with the messy gruel the Massa enjoy but a clear, well-filtered drink intended to inebriate, not to nourish.

White sorghums are the most valued by the Toupouri and are included in the ritual cycle. As among the other groups, the religious festivals and family and village celebrations provide opportunities for the ostentatious display of food. Here the traditional priest is a wealthy man, reaping benefits from a centralized political power and surrounded by a circle of assistants and dignitaries, all of whom should be well-fed and plump. Each religious celebration provides the opportunity for a banquet, which the court disposes and enjoys. If the priest's assistants are not satisfied with what they receive, he loses his supremacy and may become destitute. If he loses weight or falls sick, or if his family becomes poor, this is interpreted as wrath from the supernatural powers, which will, in turn, also harm the Toupouri people.

The traditional chief receives tributes of goals and sheep from all the villages under his control. Here we witness the beginning of a centralized political power and a hierarchy which can also be translated in terms of good food. Among the ritual foods, one notices two special delicacies: a kind of sausage made from sheep's and goats' grease wrapped in hibiscus bark - a soft, smelly food easily eaten by the toothless elders of the village. The other specialty is a sauce made from the crushed forelegs of all the cattle slaughtered on ritual occasions throughout the year and which are, of course, at various stages of decomposition. This sauce is intended to flavor the thick millet gruel prepared for New Year celebrations.

Food hospitality is compulsory within this group. The Toupouri enjoy among their neighbors a strong reputation of epicureans. They are looked upon as being able to ingest not only decayed food, but also frogs and dogs.

The comparison of these three groups corroborates Herkovits' assumption: "The more societies are able to free themselves from subsistence and have at their disposal surpluses, the more food is used in a prestigious way." (14) Here, however, economic wealth and social structure do not yet allow for the birth of an aristocracy.

It also seems necessary to assess the role played by food in differentiating one given society from its neighbors. Food ridicule is a very efficient criterion for tracing the geographical limits of an ethnic group; it may even be sharper than language. The xenophobia displayed with regard to food habits may either indicate complete heterogeneity or conceal cultural convergence and implicit subordination between different groups. This is the situation illustrated by the Moussey, Massa, and Toupouri, the latter figuring at the top of the hierarchy. From 1960 to 1975 these three groups have been exposed to two types of influence: The negro-urban, strongly marked by western society which, on technical grounds, has mainly resulted in the introduction of industrial crops, cotton and rice, neither of which has so far been able to bring these groups into market economy. The average income is less than fifty dollars per year. Cotton has diminished the acreage of food crops. Rice, which has a bad image since wild rice (*Oryza barthii*) was used traditionally as a shortage crop, is not yet very well accepted by the Massa who are supposed to grow it. These groups traditionally grow tubers such as cassava and have spread their use in this area, which is a positive feature since they can be harvested all year round.

Western models are also transmitted by the white-collar minorities from the southern areas of Cameroun, working mostly in the administration.

Globally, food resources are becoming scarce. Fish catches are less abundant, mostly due to the use of nylon nets with close mesh and the use of pesticides for fishing purposes. The increasing density of the human and cattle populations makes it more and more difficult to keep sedentary cattle. Nevertheless, cow husbandry is extending to the Moussey as a prestigious activity. In the whole area it is interesting to note that while cows, sheep, goats, and poultry are eaten exclusively on traditional occasions, Barbary duck, recently introduced, remains outside the ritual field and is widely used as a profane meat readily offered to guests.

The monetary income is too low and too unevenly spread throughout the year to allow for the regular use of modern foods. Only bread, tinned sardines, sugar, salt, and tomato sauce enter into the diet.

The picture is not the same as regards alcoholic drinks; their consumption is booming: sorghum beer and alcohol, manufactured beer, imported wine, and alcohol, in order of preference and prestige. In this connection, the behavior of the salaried workers has strengthened a traditional trend. Nowadays, drunkenness has become the symbol of monetary affluence in the non-Muslim section of the population. Beer brewing and distillation of alcohol from local grain are the most efficient ways for a housewife to earn money. This tendency should, of course, receive more attention from the government.

The second type of influence is exerted by the Fulani, who established their military domination during the nineteenth century and ever since have proven very efficient in the economic field and clever in the political one. Most of the features of Fulani culture - food production, domestic life, housing, dress and horse array, Muslim religion as well as food habits - are strongly valued by the neighboring animist groups. If one adds the fact that this group is strongly backed by the central government, it is not surprising that it has become the leading group in North Cameroun and is copied in every field. All the non-Muslim groups, including those with which we are dealing, are now deeply influenced by the Fulani and have adopted their language, religion, and material culture.

The Moussey, Massa, and Toupouri have long ago implicitly admitted their submission to the Fulani, whom they asked for supplies during the starvation periods and with whom they exchanged slaves for cereals. Today they still come to work in their white sorghum fields in order to earn money.

A Massa myth demonstrates their inferiority complex. "One day, Laona, the supreme god, created man, put him on the earth and provided him with a wife. She bore three children: the first was a Fulani, then came twins - the white man and the Massa. One day God decided to come down to earth and visit the children, but the woman was afraid and took the Massa child and hid him in a large clay urn. God spoke to the first two children. The Fulani received a bow and arrow and everything necessary for warfare. The white man received a gun. As for the third child, God said that, as his mother had hidden him, he would receive all that she was able to give him. This is why the Massa knows nothing and tills the field with a hoe."

Most of the animist groups hired by the Fulani learned their cultivation techniques for pricked-out white sorghum (15, 16, 17, 18). As we have seen, the Toupouri adopted it first and probably acted as secondary trendsetters to the Massa and Moussey. Among the former, to begin with, it comes under a very severe magico-religious prohibition, probably a defense reaction from the traditional Massa culture: "He who dared plant white sorghum would die before the ripening of the crop."

Today pricked white sorghum has spread to most of the groups, which represents a progress on technical grounds, since it can be cultivated on flooded areas and has a good yield. Favoring this new crop results in shifting priorities at farm level. Traditional red sorghum (Sorghum caudatum) now appears to be backward, bitter-tasting, and dirty. Compulsory schooling and college life have strongly reinforced this attitude. The Muslim and animist white sorghum eaters insult the Massa students by calling them "red sorghum eaters." In fact, the mere act of serving in a college refectory a meal based on the traditional cereal would immediately result in a students' strike in Northern Cameroun. It appears very likely from the above that the red sorghum staple of about half a million persons in Northern Cameroun, influenced by the Islamic and Fulani cultures, may become a white grain cereal. This tendency has a positive effect on the Massa as it modifies their attitude towards rice since it is white and they accept it more readily as they see the Muslim and the white-collar workers using it.

However, in the field of food production, the consequences need to be closely evaluated. The red sorghums can be planted on rather light ground and have a reasonable yield. Also, they are an early crop. The white sorghums need heavier soil and are late crops. This new tendency may result in extending the shortage period by about two months. Red sorghums can be consumed by the end of August, white ones seldom before October. It will be necessary to take action to increase the consumption of rice, which will have to be purchased and is, therefore, not a very good solution due to the low income. It may also be advisable to develop new varieties of sorghum, for instance those of the _caudatum_ series which produce light colored flour, or to improve white sorghum varieties adapted to light soil and which can be harvested early. A third possibility can be envisaged: the selection of a new variety of bullrush millet (_Pennisetum_). This species already gives white flour, can be planted on light soil, but has a low yield which needs to be increased.

It would be possible to show other ways in which the adoption of the Fulani pattern of life will modify the food habits of the animist groups of Northern Cameroun. The case we have dealt with is striking; in other instances it would be more difficult to unveil such dynamic trend-setters as the Fulani. In rural areas it is probably easier to do so than in urban zones. In cities the situation is much more complicated and the increasing weight of income in determining food choice makes it difficult to sort out the influence of prestige on food choices. It is, nevertheless, probably possible to do better than merely extrapolating stereotypes drawn from the observation of western urban societies.

If the remarks made by Professor Trémolières and his group in an article (19) published a few years ago can be extended to developing countries, it would appear that, since food cravings are satisfied quantitatively and qualitatively in affluent societies, prestige considerations shift from food to other areas: housing, cars, leisure, or cultural activities.

At this point one wonders whether food fads can be classified as leisure or cultural activities (20) and, if so, whether they are not surreptitiously reintroducing food as one of the criteria for social status differentiation.

REFERENCES

1. Oliver R: La longue marche des Cuistots. In: A Table, les Francais! Historia Hors Série 42:3, 1975.

2. Mauss M: Essai sur le don, forme et raison de l'change dans les sociéties archaïques. Année Sociol, 2° série, I, 1923-24.

3. Piddocke S: In Vayde AP (ed): The Potlatch System of the Southern Swakiutl - a New Perspective in Environment and Cultural Behavior. New York, Natural History Press, 1969, pp 130-156.

4. Indian Council of Medical Research: Annual Report, The Nutrition Council Research Laboratory 7:40-41. Hyderabad, Deccan, 1962.

5. Mersadier Y: Budgets familiaux africains. Et. Sénégalaises 7:82.

6. deGarine I: Usages alimentaires au Sénégal. Cahiers d'Etudes Africains, p 255.

7. de la Reynière G: Almanach des Gourmands. P. Waeffe, 1968, pp 8, 10, 11.

8. Brillat-Savarin: Physiologie du Gôut. Paris, 1885, T 1, pp 1, 158, 169.

9. Jelliffe DB: Parallel food classification in developing and industrialized countries. Am J Clin Nutr 20: 279-280, 1967.

10. Cros J, Toury J, Giorgi R: Enquêtes alimentaires a Khombole (Sénégal). Bull Inst Nat Hyg 19:650, 1964.

11. Makarius R, Makarius L: L'Origine de l'Exogamie et du Totemisme. Paris, 1961.

12. Bartlett HG: Innovation, the Basis of Cultural Change, p 51.

13. Masseyeff R, Cambon A, Bergeret R: Enquête sur l'alimentation au Cameroun III Golompoui. Yaoundé, IRCAM, 1959, p 16.

14. Herkovits MJ: Man and His Works. New York, Knopf, 1950, p 284.

15. le Conte J: Enquête sur les sorghos et mils du Tchad. IRAT, 1965.

16. _________: Note sur les sorghos du Nord Cameroun. IRAT, 1965.

17. Marathee JP: Etude concernant la prospection sorgho de trois départements du Nord Cameroun (Margui, Wandala, Dianare, Mayo Dania.)

18. Guillard J: Golompoui, Nord Cameroun. Paris, La Haye, 1965, pp 226-302.

19. Claudian J, Serville Y, Trémolières F: Enquête sur les facteurs de choix des aliments. Bull de l'INSERM 24:1346, 1969.

20. Kühnau J: Food cultism and nutrition quackery in Germany. In Food Cultism and Nutrition Quackery Symposia of the Swedish Nutrition Foundation, VIII. Upsala, 1970.

FOOD FADDISMS

Fredrick J. Stare, M. D.

Professor of Nutrition
Harvard School of Public Health
Boston, Massachusetts

INTRODUCTION

For many centuries, people have attributed special characteristics to certain foods. Garlic was long touted as an essential food for physical strength, cabbage was thought by some to have mystical powers to cure illness, truffles were generally agreed to assure great sexual potency, *ad infinitum*. These early food prejudices and practices have continued over the years, sometimes with slight modifications, sometimes unchanged, but never completely dropping out of the cultural heritage.

The mystique of food quackery has evolved alongside genuine advances in nutrition. During the Middle Ages, at various points around the world, special foods were being recommended for curing certain diseases. Apparently, some people felt that the more unpalatable the food or its source, the more it would accomplish toward curing the problem. This general philosophy culminated in such delectables as ground coffee beans blended with fat! There is still an unstated but apparent attitude today among food faddists that the general appeal of the food is far less important than the wonderful things it will do for your health.

The United States, like most other countries, has had its share of quacks and faddists. Among our most famous hucksters of earlier generations may be counted Sylvester Graham (of Graham cracker fame), Horace Fletcher (the widely followed exponent of the virtues of chewing each bite 32 times, one chew for each tooth), and finally the late Bernard Macfadden (self-acclaimed king of the physical culturists). These noteworthy businessmen acquired a distinguished following over the years. Henry Ford, John D. Rockefeller, Sr., and George Bernard Shaw are but a few of the famous who have faithfully followed the notables in quackery and food faddism. An equally prominent list of contemporary individuals could readily be given. Some people have made fortunes selling pseudonutrition, many have paid considerable cash to follow the dictates of these leaders, many have paid with poor health, and some have died as a result of following the golden rainbow to health through pseudonutrition as prescribed by quacks and charlatans.

The idea that there have always been people seeking good health through food is not surprising. It is not even too amazing that some unusual and highly restrictive diets were tried as late as the nineteenth century. However, the fact that many people today still believe in virtually magical powers in some foods to the exclusion of a normal diet is difficult to accept. With the remarkable progress in understanding the role of food, and the significant discoveries resulting from the dynamic development of nutrition as a science, it is truly amazing that so many people in the twentieth century still select their diets on the basis of emotion and mysticism instead of scientific fact.

THE HEALTH FOOD BONANZA

There are several reasons why food faddism and quackery continue to flourish. Certainly the scientific knowledge of nutrition is ample to enable the normal individual to be well fed. The difficulty begins to develop at the point of transmission of that knowledge from the scientific nutrition laboratory to the training of physicians, allied health workers, and ultimately to the individual consumer. Until all people have some accurate and practical nutrition information regularly included as a part of their general education, beginning with the earliest experiences in family life and school, there will be large numbers who lack the knowledge required to shield themselves from the barrage of misinformation being leveled at the public today.

The natural human desire for an easy solution to a problem sets the stage for promoting all sorts of pills and diets for effortless weight loss. Hypochondriacal tendencies are fertile territory for breeding health worries that can be allayed by eating or drinking a special food preparation. The gullible nature of some people who will subscribe to any theory with a scientific sound brings still others into the world of the faddist. These psychological undertones provide the foundation for health business enterprise, thus attracting imaginative and enthusiastic, if not altogether altruistic, entrepreneurs. Any business so closely attuned to the public's concern for health and long life and so intimately tied to John Q. Public's wallet is virtually assured of continued success, at least until nutrition education becomes far more effective than it is at present.

A visit to a health food store is an important part of studying this aspect of nutrition. For the person who has never journeyed over the threshhold of such an establishment, a new experience awaits. Some of the more plush stores boast a juice and nutrition counter featuring such delectables as celery juice. The fresh produce section houses fruits and vegetables (frequently scarred with worm holes and other indications of insect interest) that would be hard pressed to attract a buyer in a regular grocery store, but that will sell to the food faddist because they are clearly marked "organically grown." A prominent display of vitamin and mineral capsules is almost a certainty, for these are high-profit items. Various elixirs, herbs, oils, sweeteners, and other fascinating food products await the interested customer. No visit would be complete without a careful survey of the literature available in the store. The choices range from leaflets extolling the amazing return to youth one can experience by drinking a tea brewed with desert weeds to books assuring the reader that he can lose weight while eating all that he wants. Such a store is directly keyed to arouse people's concern over their health and then to capitalize on this apprehension by offering the solution to a problem that the buyer might not even have thought of before entering the store.

A comparison of prices between a chain store supermarket and a health food store quickly reveals that many of the items in the health food store carry higher price tags than those in the supermarket, despite the fact that the quality may not be as good. The extra expense for health foods does not stop there. Health food store patrons are constantly urged to buy dietary supplements and tonics. Such items usually are fairly expensive, an expense that need not be incurred at all if one is in normal health and follows a well-balanced, varied diet.

All the foods needed to provide ample quantities of all the nutrients required by man are readily available in regular stores at competitive prices. In fact, no vitamin or mineral supplement is needed by normal, healthy people who are eating a balanced diet.

Aside from the extra expense of eating foods from the health food store, there is also the fact that food faddism can present a health hazard. Persons with physical ailments that need immediate medical attention may delay seeking medical help while they try to treat themselves with a special diet or product from a health food store. Such postponement of needed medical attention may introduce unnecessary complications and hazards.

There is another side to the coin that also must be presented. Many foods available in health food stores are highly nourishing and do provide considerable interest and flavor appeal to a menu. Bread made with stone-ground whole wheat flour, for example, is a food to be enjoyed by all. To condemn foods simply because they are highly touted by food faddists is narrow minded. Many foods embraced by the faddists add an additional flair to meals and are enjoyable, despite the fact that they are not the panacea to cure all ills. Nourishing foods can be purchased in grocery stores and health food shops. If the shopper in a health food store buys some foods there because he likes their taste and texture and wants them as a part of a well-rounded diet, with price not being of any particular concern to him, obviously the purchase options are his to make.

DEFINITIONS

It is always helpful to know what one is talking about when discussing nutritional faddism and nonsense, and hence if we are to discuss nutritional faddism we ought to define it. The best place I know of in the United States for definitions is Webster's Unabridged Dictionary. A fad is defined as a "silly thing, followed for a time with an exaggerated zeal." So nutritional faddism would be "silly nutrition followed with an exaggerated zeal."

Quacks and charlatans are two terms I have not defined until now, but they are terms we frequently use in referring to the vendors of food faddism and nutritional and health nonsense. A quack is defined as "a boastful pretender to knowledge or ability." Thus, were I to boast about my knowledge of radio, television, or many other things where I really have very little knowledge, I would be a quack in those areas. An individual boasting about his knowledge of food and nutrition, and with really very little or no training in biology, chemistry, or nutrition is a nutritional quack.

A charlatan is defined as "one who boasts much in public, makes unwarrantable pretensions, is usually a vendor of remedies, and is a pretender to knowledge or ability." A charlatan is thus a quack with something to sell - a worthless tonic, foods with unusual health properties, or a series of lectures or a book on subjects for which he is not professionally qualified.

"Health foods," "organic foods," and "natural foods" are three terms currently in common usage in food faddism and they should be defined.

While our most recent unabridged dictionary contains no specific definition of "health food," it makes it obvious that "health foods" are those that promote health or are conducive to health. With that definition, we maintain that all edible foods when properly used are health foods and promote physiologic or psychologic health - regardless of whether they are purchased in a neighborhood grocery store, a supermarket, or a so-called health food store.

The dictionary is more specific for "organic," though it still doesn't mention "organic foods." The definition of "organic" that is most pertinent to food faddism reads: "of, pertaining to, involving, or grown with the use of fertilizers or pesticides of animal or vegetable origin, as distinguished from manufactured chemicals: Organic farming."

Strictly speaking, all foods are organic - protein, fat, carbohydrate, and vitamins - because they are, in the definition of the chemist, organic compounds containing carbon. The organic food enthusiasts should speak of "organically grown foods," by which they mean foods grown on soil fertilized with organic fertilizers - manures or composts of various types - and foods grown without the use of chemical pesticides.

Such foods, of course, do not have superior nutritional qualities. As we all know, the nutritional qualities of foods depend primarily on the genetic make-up of the seed. Fertilization, either organic or chemical, influences crop yield but except for minerals, not nutritional quality.

"Natural food" is not to be found in the dictionary, but in many discussions we have had with people who believe in "natural foods," their definition is identical to what we have given in the preceding paragraph for organically grown foods, but with the addition that in the processing of the foods, no substances (nutritive or non-nutritive) are added to the foods. If one adds even such common and simple substances as salt or sugar to "natural foods," they are, strictly speaking, no longer "natural foods."

We think no foods should be identified as "health foods," because as previously stated all edible foods, when properly used in a balanced diet, are conducive to physiologic or psychologic health - regardless of whether they are purchased in a so-called "health food store," an ordinary grocery or supermarket, or in a special section of the latter.

We think no foods should be singled out as "organic foods" because all foods are organic. Foods that are fertilized only with organic fertilizers, rather than manufactured chemical fertilizers, and with which no chemical pesticides have been used in the growing and no chemicals of any kind added during the processing and preparation (food additives), might be identified as "organically grown foods." However, even this terminology is not correct because much of any soil is composed of inorganic substances that are necessary for the growth of plants.

We think that the term "natural foods" should be abandoned because all foods are natural or are manufactured from natural foods. However, many individual nutrients - for example, most of the vitamins - may be either of natural origin or manufactured synthetically. However, these individual nutrients are not foods; they are nutrients. Thus, vitamin C (ascorbic acid) occurs in citrus fruits - foods - or it can be manufactured chemically as a nutrient, not a food, and then added to foods that may be low or lacking in it. If this is done, it should be indicated on the label, and it usually is. Vitamin C and other nutrients - whether they occur in foods or are manufactured and added to foods - are identical chemically and biologically.

WHY SO MUCH FOOD FADDISM AND QUACKERY

Before closing I should like to make a few comments about quackery that apply to all types, not just nutritional.

There are strong psychological aspects to the public's susceptibility to food faddism and quackery.

Fear is a basic cause of vulnerability to quackery. Fear of illness, physical or mental incapacitation, weakness, and death returns us to the childish condition of wanting reassurance and strength from an uncritical adult who promises safety and well-being. This is when the quack steps in and takes over.

Much of modern medical and dental practice is, unfortunately, characterized by a brief, impersonal relationship between doctor and patient. We see much pain, disease, and suffering, all of it in detail and with unavoidable comparison to any similar cases. Hence, a patient's particular complaint is put by the physician in its proper perspective as to its severity and need of treatment. Perhaps, too often, the proper prescription is something as unimpressive as a "couple of aspirin." But this is not enough to satisfy an individual who considers his complaints severe and deserving of sympathy and serious treatment. So he turns to an advertisement which expresses an understanding of his pain, his need for a new and simple treatment. This is when the quack and charlatan step in to offer him sympathy, packaged as cleverly as the nostrums they sell.

There are those who turn to quackery *in extremis*: Intelligent, well-educated people who have incurable disease may seize upon any promise of hope, and hope is what the quack offers in abundance. These victims are easily gulled by the quick, easy, and absolute relief the quack offers.

The vast majority of people are not trained in medicine or science and many of them have varying feelings of anti-medicine and anti-science - and so does the quack and charlatan - but they all have a certain ambivalence about this. Most people will readily acknowledge that physicians know something about medicine, but few persons place great confidence in those doctors they know or those in their own community. The charlatan plays on this ambivalence by running down the local physicians and medical societies, yet saying that a famous physician, usually a long, long way off, has used the proffered remedy or diet with great success.

Debunking of food faddism and quackery's false and extravagant claims is a challenge to all in the health profession. It is also a challenge to those in the business of communicating with the public via newspapers, magazines, radio, and TV, regardless of whether the communication is categorized as news, editorial, letters to the editor, or paid advertising.

EATING - SAFELY - THROUGH THE EIGHTIES

Food and Disease

We've always been concerned about the possible relationship between food and disease, and it is highly likely that we will come to be even more concerned in the future. But the focus of that interest will not be the "artificial" vs. "natural" dichotomy as amplified by the food faddists, quacks, and charlatans. Rather it is probable that new information about both the toxicity and health-promoting effects of <u>all</u> types of food products - no matter what their origin - will give us a better understanding of the nature of the link between eating patterns and chronic disease. The food technology research of the late 1970's and 1980's will be focused on two related goals: the need to provide an adequate, healthy and inexpensive food supply for a rapidly expanding population - and the desire to develop the least hazardous, most health-promoting human diet.

FOOD IN A GROWING WORLD

The world population is now over the four billion mark and is increasing at a rate which will cause the population to double about once every 35 years. Demographers have presented us with such staggering figures that we tend to block out the whole subject, but the reality of rapidly increasing humanity is with us.

We tend to forget that the preoccupation of most of the world is not about additives and their safety, as the faddists would have us believe, but about getting enough food to live a healthy life. In global terms in the 1950-1970 period, with the widespread use of fertilizers and pesticides, food production increased at a faster rate than did population growth. (The world production of grains almost doubled while the population increased by less than 50 percent.) But the situation has deteriorated in recent years, in large part because of droughts and general poor weather and because of higher birth rates. Starvation in the world is again becoming a leading cause of death - particularly among young children.

We need more food - both in the developed as well as the developing world. Intensive efforts are necessary both for increasing crop yield in each agricultural acre and for reducing food deterioration and wastage. In approaching this problem, we might consider three alternatives.

First is the return to "natural, organic, pesticide and additive-free living," and the suppression of research which might develop new "chemicals" for food. Obviously, if we are now having problems in developing countries in meeting food needs and in developed countries being able to afford the food which is available, abandoning the techniques of food technology which have taken generations to develop would only complicate the situation. There is no time for food faddism in terms of world food problems.

Second, we could look for new ways to protect and preserve food, for instance, techniques which do not involve the ingestion or digestion of food additives and pesticides. Indeed a company in California is currently focusing its attention on the possibility of developing indigestible additives. Their primary concept is based upon the physiological fact that digestion is largely a process whereby food is broken down into its molecular components, which are then small enough to pass through the intestinal wall. Molecular size is, generally speaking, the only barrier to intestinal absorption. So, if the additives could be altered in some way which would prevent their absorption, and still allow their functions to be performed, there is, at least theoretically, the possibility that current additives could be used in a manner which would not cause alarm among even the most dedicated faddists. But that's in the future, if at all.

Third, we could move ahead in all aspects of food technology research, looking for new sources of foods, new and safe additives which could further reduce spoilage, and lead to the creation of new, highly nutritious, and inexpensive food supplements which would be culturally acceptable in areas where food supplies are limited. But a prerequisite of this type of rational approach is a trust in the "scientific system" and the acceptance of the fact that our current food research protocol has safety as its number one priority. To move in the direction of increasing and improving our food supply through scientific knowledge, we must have confidence in our academic scientists, the food industry, and the regulatory agencies which govern them. Right now, however, confidence in these groups seems to be waning.

Of course, this disenchantment with science is nothing new. We've long known that while technological abundance in any area initially elicits feelings of happiness and gratitude (when, for example, sanitary water, sewage disposal, and pasteurized milk became available people responded positively; they could see the direct impact these public health measures were having on themselves and their families), the romance soon fades and is replaced by discontentment and suspicion. We must return to the reality of life and acknowledge that we do owe a great deal to the scientific advancements of the last seventy or so years - and that our hope for the future is based on the assumption that those advances will continue. Until, however, this feeling of confidence is reestablished, food research will be inhibited. Industries are understandably not eager to invest large revenues in projects which could be terminated by even a hint of a health hazard as, for example, at the "drop of a rat," through some animal study using dosages completely unrealistic to the use of the substance in the diets of man or even giving the substances in large amounts intravenously.

Good nutrition for young and old is obtained quite simply by eating a variety of foods - a good source of protein which, for most of us, is usually meat, chicken, or fish; some kind of a milk product such as milk or cheese; fruits and vegetables of any kind (but with variety!) and some foods made out of cereals of any kind. In the United States this variety is referred to as the Basic Four Food Groups. And above all, again for most of us, adjusting our total caloric intake, and that includes the calories from alcoholic beverages, to our caloric output so that proper growth is obtained in children; and proper weight is maintained in the adult. Again for most of us and our friends that usually means to eat and drink less and be more active physically.

CONCLUSION

As I mentioned in the beginning of this paper, food faddism has been around a long, long time and in all parts of the world. It is still strong, perhaps even growing. I wish to call your attention to three recent references on the subject.

The Journal of the American Medical Association in one of its recent issues (August 11, 1975) has an interesting editorial by Dr. Philip White, Director of the Department of Foods and Nutrition of the American Medical Association. It is entitled "Megavitamin This and Megavitamin That."

In the same issue of that journal, Dr. Thomas H. Jukes of the Division of Medical Physics, University of California at Berkeley, writes on the same subject - Megavitamins. Dr. Jukes concludes: "A regrettable side effect of quackery, magnified in our era of instant communication, is that extensive efforts of responsible scientists are needed to test and expose irresponsible proposals and suggestions that promise 'health' from overdosage with universal remedies. These efforts are diverted from more constructive activities, and society is the loser.

Lastly, I wish to call your attention to a new book of mine, done jointly with a former student, Dr. Elizabeth Whelan. The book was published only two weeks ago. It is titled _Panic in the Pantry_. It deals largely with food additives, what we call "Health Food Land," and food faddism. The gist of the book is: have confidence that our foods are safe, eat a varied diet, don't eat (or drink) too much, and be wary of those who continually raise questions about the safety of our foods and who continually come up with a new diet to cure the ills of mankind.

TEACHING ASPECTS

EDUCATION AND TRAINING IN NUTRITION

Hans D. Cremer, M. D.

Institute of Nutritional Science
Wilhelmstrasse 20
6300 Giessen
Federal Republic of Germany

The previous speakers have discussed problems in many areas of nutrition. We have heard about food production and the availability of food, about the interdependence of food and health, and about malnutrition which may, in part, be caused by poor eating habits and aggravated by food faddism. There is no doubt that nutrition problems are world-wide.

In the developing countries nutrition is a primary problem. There, a low intake of calories and a deficiency in various nutrients: proteins, vitamins, and minerals, result in the so-called deficiency diseases. These have almost completely disappeared from the European continent and from the United States, but in their places are nutritional problems which fall into the field of social medicine, having resulted from rapid changes in living conditions, specifically urban concentration, industrial development, and technical advances which have increased the life span. We must consider how to fight malnutrition wherever it exists, whatever its nature, and whatever its causes.

Awareness of nutritional problems led to recognition of the need for scientific research in this field. In the Western countries centers for the study of nutrition problems, institutes for specialized research, and scientific associations, both in universities and under private auspices, have been created.

This burgeoning interest in food and nutrition has spread to other disciplines including medicine, biochemistry, agriculture, and technologies related to food production. Nutrition was recognized as a separate science about twenty years ago when the International Union of Nutritional Science was created within the framework of the International Unions of Sciences. However, in Europe as well as in the United States training of teachers in nutrition and dissemination of nutritional information have not developed to the same extent as has research.

From the teaching aspect one must differentiate between nutrition training, which is mainly professional, and nutrition education, which deals with the education of the public.

Originally specialists in this field were essentially self-taught, and for a long time this seemed sufficient. However, about a decade ago, when research programs were extended and numerous forms of collaboration developed, the need arose to consider how to teach as well as who to teach.

To find ways to meet this need two international groups, the Food and Agriculture Organization of the United Nations (FAO) and the World Health Organization (WHO) organized a symposium which was held in December, 1959, at Bad Hamburg, Germany. Twenty-four European countries sent over sixty representatives to discuss problems related to both professional and vocational teaching and training in the field of nutrition. These questions were considered from the viewpoints of nutrition and applied dietetics, medicine and public health, home economics and domestic sciences, and agriculture.

One of the resultant recommendations was that a mission be sent to assess the status of nutritional education in various European countries, by discussing the problems with the government authorities and the personnel engaged in the field of nutrition training and teaching.

In 1961, when I was Chief of an Applied Nutrition Branch, I was responsible for creating such a mission. I spent a week in each of the following countries: Belgium,

France, Italy, The Netherlands, The United Kingdom, and Western Germany. In general the visits included:

-- round table discussions held with the government authorities, university members, and others directing or taking part in the study programs of various educational institutions from the university level down to the elementary school level, and

-- visits to some of these institutions, in order to obtain impressions of the differences among the countries and in the kinds and levels of education.

The situation remained static during the next decade until a new international meeting, the joint UNICEF/WHO Seminar on the Teaching of Nutrition in Mediterranean Countries, was held in Ankara in December, 1970.

NUTRITION IN MEDICAL STUDY PROGRAMS

As the relationship between nutrition and health became recognized, the need to include the study of nutrition in the curricula of medical schools became apparent.

Its importance was considered both in the general training of medical students and physicians and in medical specialization or postgraduate study programs.

A basic knowledge of the composition of foods and of metabolic processes is part of some curricula within the general Doctor of Medicine program and/or within the classical basic science programs of physiological chemistry, biochemistry, and physiology during the years of preclinical training. However, the number of areas to be covered, both in physiology and in biochemistry, have become so extensive that it is completely impossible, within the lectures, to deal with all of them. Therefore, whether or not nutrition is taught seems to depend very much on the inclination of the professor in physiology or biochemistry. Later, in the clinical fields, knowledge of nutrition is essential in its relationship to human pathology, in hygiene, and in the clinical internship training in hospitals, in therapeutic diets and the feeding of patients.

In pediatrics, the area of nutrition has long been of primary importance, but the concepts of food and nutrition are scattered sporadically under varied names and with no attempt at integration within the classical study program.

Though this form of instruction for a while seemed adequate, some faculties now feel the situation to be unsatisfactory, because these subjects are all too often taught by professors who are not nutritionists themselves, and, therefore, have no personal interest teaching in it.

Those members of medical faculties, who severely criticized the prevailing classical views, deplore the fact that the student has to consider the changes in the diet necessary for the treatment of sick patients before he has been taught what the constituents of a normal human diet are, the various factors involved in determining what that diet should be, or the reasons that the dietary regime of peoples is made up differently in different parts of the world.

Many professors have expressed the opinion that it would be advisable to deal with these questions either in courses in hygiene or in social medicine; but it has also been pointed out that this could not be done satisfactorily unless the professors of those departments, who are not usually nutritionists, would allow a number of lessons in their courses to be given by qualified specialists in the field of human nutrition. In almost all countries the future general practitioner learns little about the practical problems of dietetics that arise in his professional activity, either with respect to the normal or to the sick patient. He is left to fill in the gaps in this aspect of his education the best way he can. In order to guarantee that the medical doctor has some knowledge about nutrition, the teaching of nutrition must become part of the medical curriculum.

In the Federal Republic of Germany a federal regulation for the medical examination ("Approbations-Ordnung") was introduced in October, 1970. A part of this examination consists of answering a number of multiple choice questions on nutrition. I think this is the first instance where a country has required its students in medicine to have a basic knowledge in nutrition. This development will, no doubt, increase the amount of attention given to nutrition teaching, both by the teaching staff and by the students.

The IUNS has created a Committee on Nutrition Education in Medical Faculties. This committee intends to submit evidence concerning the urgent need to include the teaching of nutrition in the education of medical undergraduates and postgraduates in schools of medicine all over the world and wishes to make practical proposals presented within its conclusions.

The committee has examined available reports on programs of nutrition teaching in medical faculties and has found - comparing the situation between the two international congresses of nutrition (1966 in Hamburg and 1972 in Mexico) - certain interesting developments that have occurred which bode well for the future. For example, the Nutrition Division of the Department of Community Medicine in the Mount Sinai School of Medicine in New York City has a comprehensive program of nutrition for its medical students. In Great Britain, Cambridge University is introducing a nutrition teaching program in its clinical curriculum, which includes a pregraduate clinical course lasting two years and a postgraduate diploma course in nutrition directed by a committee representing the faculties of medicine, biology, veterinary medicine, agriculture, biochemistry, physiology, and a number of research institutes.

In spite of these first steps the IUNS committee states that, on the whole, education in nutrition in most schools of medicine is not adequate. Therefore, the members of this committee are recommending that medical students and medical graduates should be kept informed of the advances in nutritional sciences which have occurred in the recent years, since it considers it an essential duty of the medical profession to be contributors in preventive public health and nutritional measures, as well as to use current nutritional knowledge in the treatment of nutritionally dependent disease. The importance of nutrition in conditions of physiological stress, e.g., in growth, pregnancy and lactation, or after injury, as well as in diseases with a dietary component (obesity, heart disease), is of primary interest to the medical profession.

The Council on Foods and Nutrition of the American Medical Association has stated that "in general, medical education and medical practice have not kept abreast of the tremendous advances in nutritional knowledge. A recent survey on nutrition teaching in medical schools indicated that there is inadequate recognition, support, and attention given to this subject in medical schools."

Before discussing nutritional training outside the field of medicine, I should like to give the text of the recommendations of the two international congresses of nutrition on this subject. That of the WHO Seminar of 1970 was the more general. It follows:

-- Nutrition teaching in the medical faculties of Mediterranean countries should be improved.

-- The most practical solution is to include nutrition teaching in that of the relevant disciplines.

-- To effect such integration a coordinator is essential. Medical faculties should have a professor specializing in nutrition or a professor of nutrition science who is responsible for such integration and coordination.

The recommendations of the IUNS Committee are:

-- The physiological and biochemical bases of nutrition should be taught in the preclinical years.

-- The pathology and therapy of nutritionally induced disease should be taught in the clinical years.

-- The public health and community medicine aspects of nutrition should be an integral component of the medical curriculum.

-- Clinical nutrition and preventive or therapeutic dietetics should be presented to medical students in such a manner that they can make use of this knowledge in medical practice.

-- Active efforts should be made to support postgraduate education which would keep the practicing physician informed of advances in nutritional knowledge and would also provide information of a more specialized kind, e.g., in the form of symposia or courses on nutritional problems in their own country or on nutritional medical practice in developing or tropical countries.

-- A medical school should place authority in an individual or committee, or preferably found a Chair of Nutrition, with responsibility to propose an integrated and full teaching program in nutrition which would cover the efforts of various departments.

-- Since the recommendations cited above are necessarily of a long-term nature and since they are unlikely to be implemented fully in the near future, despite their urgency and desirability, the Committee urges the Council of the IUNS to take measures which would be in partial fulfillment of our aims.

The Committee, therefore, further recommends that the Council of the IUNS should sponsor, facilitate, and arrange financial support of about five to six postgraduate students, to be called IUNS Fellows, to attend special courses in nutrition (diploma, etc.) so that these Fellows, who would come from both industrialized and developing countries, would become themselves teachers in nutritional sciences as medical nutritionists, acceptable to medical schools. An approved list of training centers willing to perform such a task should be made and financial backing found from outside sources. Every effort would have to be made by both the granting body, e.g., IUNS, and the medical school to assure a suitable position in the medical faculty for the returning scholar. This would ensure a continuous succession of enthusiastic, young, well-trained nutritionists who would enhance the nutrition education in their respective medical schools.

PARAMEDICAL STAFF CONCERNED WITH NUTRITION PROBLEMS

In addition to the physicians themselves the paramedical staff should be used to assist them in teaching aspects, to help to combat malnutrition, and to ensure that people are properly fed. Therefore, a certain degree of nutrition training should be included in the curricula of the different categories of paramedical staff.

Here we must differentiate between two groups:

- -- personnel for whom nutrition is the main concern - dieticians and/or nutritionists who should be grouped in these categories:
    - -- hospital dieticians
    - -- public health (community) dieticians
    - -- public health nutrition workers
- -- personnel who are not mainly concerned with nutrition - nurses, midwives, pharmacists, and, to some degree, medical technicians.

I should like to go into more detail concerning the duties of hospital and public health dieticians. Hospital dieticians should have, besides their work in general, catering and in dietetic therapy, some teaching responsibility. This need arises when patients are admitted to the hospital and require dietary consultation.

In some respects the hospital dietician also is responsible for teaching other hospital paramedical staff, and when needed, patients' families.

The duties of public health dieticians and/or public health nutrition workers are broader and more varied, especially in their teaching obligations. Their activities include advising governments and public health departments, organizing and supervising nutrition of different communities, therapeutic functions within public health institutions, and organizing and supervising public health programs. All of these activities may include teaching aspects, and, therefore, teaching obligations.

The IUNS Committee on Nutrition Education and Training in schools of home economics, nutrition, dietetics, and allied health professions published a report in 1972 reviewing the educational preparation and types of employment of dieticians and nutritionists throughout the world. The review did not give details of the course content of the studies nor a description of the specific functions performed within the services in which the graduates work. It assumed, however, that the schools provide adequate preparation for the types of activities for which the graduates would have responsibility and the kind of work the employment market demands.

Information was obtained from 46 of the 125 member states of the WHO. This study identified those countries with schools which prepare specialized personnel in nutrition and dietetics, the type of school offering these courses, i.e., universities, hospital schools, technical or professional schools, schools of home economics, and others. It further provided some information on admission requirements, length of the total education and training period, whether a certificate, diploma, or degree is awarded, and the type of institution or service in which the graduates are employed (2).

NUTRITION AS A SUBJECT AND AS A SCIENTIFIC DISCIPLINE

When the FAO/WHO mission visited European countries in 1961, they found that nutrition did not appear by name as a subject or as a scientific discipline with a special department of its own in institutions of higher learning as did, for example, the chemical or biological sciences. Since then the situation has changed or is changing in many countries. Today in some countries there are both schools of nutrition and schools of home economics. The conditions in my own university are quite special, so I should like to describe them.

In 1956 the University of Giessen created the "Institut fur Ernährungslehre," i.e., Institute for teaching Nutrition. In practice this was a research institute with an interfaculty structure supported by the departments of agriculture, medicine, veterinary science, hygiene, and pure science. It worked as a research and teaching center where only a few members of the scientific staff had been trained in medicine. For purely administrative reasons it was considered necessary for such an interfaculty body to be attached to one particular faculty. In this instance, this institute was first attached to the medical faculty of the University of Giessen. However, there developed confusion relative to awarding degrees in nutrition by a school of medicine, so a new four-year course was organized under the auspices of the faculty of agriculture, with the collaboration of the faculties of medicine, veterinary medicine, and science. It terminated in the degree of Household Economics (domestic science) and Nutrition, with specialization in either of these two fields.

A few years ago in the state of Hesse a new university law came into effect which specified a new organization within the university to join related chairs and/or departments into "Fachbereiche" (subject areas). The two subject areas of concern here were Nutrition and Household and Food Economy.

The responsibility for teaching in nutrition was borne by two Fachbereiche. Now the University of Giessen is offering nutrition as a subject, a scientific discipline as such. There are similar examples in other European countries and in the United States.

NUTRITION AS A SUBJECT WITHIN OTHER FIELDS

Since we know how important correct and well balanced nutrition is for the well-being of man, not only specialists and physicians should be instructed in it; but principles of good food and correct nutrition should be taught to members of all fields who are concerned with the production and distribution of food and the food economy. Thus, sociologists, psychologists, teachers and even politicians should have sufficient knowledge of nutrition that they can understand and interpret its aspects whenever and wherever they are relevant within their fields to educate the consumer.

EDUCATION IN FOOD AND NUTRITION

The question as to whether it is worthwhile to put much effort into nutrition education of the public, i.e., of lay people, can be decided only after having studied the possibilities of improving the feeding in the world through education. Being convinced that nutrition education was urgently needed and would be effective, Dr. Sen, Director-General of FAO in 1959, launched the Freedom from Hunger Campaign with the following objectives:

-- to make people aware of the problems of hunger and malnutrition

-- to show the consequences in human suffering of malnutrition

-- to promote the activities of national and international scales for nutritional improvement.

The Freedom from Hunger Campaign was concerned with nutrition education in the broad sense that it would teach many people to make better food choices and encourage many housewives and mothers to feed their families more properly. It sought to show the consequences of poor nutrition for health and efficiency, and the need for certain nutrients, i.e., the correct composition of a good diet.

The importance of nutrition education was strongly stressed by Dr. Autret, former Director of the Nutrition Division, FAO, in his foreword to the FAO brochure, "Learning Better Nutrition." He says:

> Ignorance is the ally of hunger. Together with poverty, which it often accompanies, it is basically responsible for virtually every case of malnutrition; and in countries where food supplies are inadequate, existing resources are generally badly utilized. Many a case of kwashiorkor could be prevented if mothers knew how to make the best use of what food there is; and the food shortages of a number of countries could be overcome if farmers were able to produce more efficiently. Educating both producer and consumer, therefore, is fundamental to any course of action in improving the state of nutrition of a population and ensuring that families spend their money wisely.

Dr. Autret made it clear that an important part of the nutrition gap is lack of information. The same ideas appeared in the declarations of two famous specialists in nutrition:

-- Cicely Williams, the famous British physician, who first described protein malnutrition in Africa and coined its name, <u>kwashiorkor</u>, agreed that in most developing countries malnutrition is caused by poverty - not <u>economic</u> poverty but a <u>poverty</u> of <u>knowledge</u> of the nutritional needs of a child.

-- Dr. Likimani, when Chief Public Health Officer in the government of Kenya, pointed out that "practically every case of malnutrition is due to ignorance, and only some are due to ignorance combined with poverty."

Prejudices and taboos are hampering correct nutrition in many developing countries. The best known example is the prohibition of eating the meat of cows in India. In many African countries the people, particularly girls and women, are forbidden to eat fish, eggs, and certain kinds of meat. Some foods of high protein value are forbidden during pregnancy. The disadvantages of such taboos are obvious.

A clear example of nutritional ignorance is the limited consumption of legumes. Different kinds of legumes are important sources of protein in many developing countries; but weaning infants are not fed peas and beans for fear that these foods will cause flatulence and illness. They are commonly eaten only by adults, though the protein requirements for infants are relatively much higher than those for older people. There are many kinds of legumes which are not tolerated well by small children, but an appropriate test can establish which kinds are suitable for children. My associate, Rolf Korte, who is now Nutrition Advisor of the government of New Guinea, has learned in Werner Jaffe's laboratory in Caracas to carry out such special tests for digestibility and wholesomeness of legumes and has applied this knowledge in Africa.

Another example of inadequate knowledge in nutrition is the vitamin A deficiency in Indonesia. Fruits and vegetables, rich in carotene, are commonly available even to low income families. Young children, especially boys, however, seldom are fed vegetables, since vegetables are not regarded as suitable food for young men.

Nutrition education is an important consideration as an element in overall nutrition strategy because it acquaints people with the value of resources already available to them, and persuades them to modify existing practices.

However, providing people with information does not automatically mean they are applying their new knowledge. An evaluation of the education programs may have been neglected on the assumption that nutrition education does not need an evaluation; or it may have been scheduled to take place at the end of a nutrition education campaign, and many projects never reach a terminal point. How important evaluation can be, even if its outcome is disappointing, is shown in an example from India. A dozen years ago, when I was Chief of the Applied Nutrition Branch of FAO, one of my duties was to start the Applied Nutrition Programs in India. After twelve years of practice, in the early seventies, Indian scientists started to evaluate some programs and, as Alan Berg reported in his recent book, The Nutrition Factor, in many cases they did not find any encouraging results: one could not detect any significant differences in general dietary practices, "particularly with respect to nutritionally desirable commodities which are promoted under the program." Knowledge and understanding of nutritional needs was no greater in villages within the program than in those outside the program. The general failure of the program was attributed more to "the conceptualization behind the program," not so much to a uselessness of nutrition education programs in general. Apparently we have to learn more about how to adapt special education methods to the special needs of people or to the special situation in a given country.

We should search for better ways and means of accomplishing effective nutrition education. Nutrition education activities in many developing countries have been disappointing and have raised the question of whether food habits can be changed by education. Recent experiments in many parts of the world and observations of coworkers in Thailand have shown that existing food habits are not completely immutable. Foods that were totally foreign to many people are now regarded as dietary staples, or will be eaten at least by many people. Themes for nutrition education include better nutrition of children and encouragement of breast feeding. Other examples are provision of food supplements to nursing infants at the appropriate age, avoidance of waste, and better distribution within the family of foods which are nutritious and adapted to the family food budget.

However, people must see a reason to change their habits. This means the education campaign must identify the problem, engage emotions, explain the need, and demonstrate the results. Krishnaswami, an Indian friend of mine, has published a large book about his experiences in nutrition education in India. I am going to mention some of his points:

Potential causes of hidden resistance to change must be identified so the message can be designed to address such obstacles as economic resistance, ("I can't afford it,") and social status resistance ("everyone else eats this way,") or ("we have been eating this way for generations.")

Since malnutrition is a mass problem, nutrition education probably ought to be channeled through the mass media, instead of face to face contact that has dominated nutrition education up to the present, but the situation in many developing countries hampers communication via the mass media. Many, often most, adults in developing countries are illiterates. In some countries language differences raise additional problems. India has 14 basic languages, many as different from each other as German is from Japanese. Some African countries have more than a hundred special languages.

Another difficulty is transportation and frequency of mass communications. In India, for example, newspapers reach only one percent of the population. Even television is not a suitable instrument of mass communication in many developing countries. From the approximately 300 million television sets in the world, only about five percent are found in South Asia, Africa, and the Arabic countries of the Middle East altogether. Nevertheless, commercial advertising and sales promotion have shown that by use of mass communication some people can be reached.

Up to now, nutrition education does not appear to have brought about large scale changes in eating habits. If nutrition education is to remain a part of our nutrition strategy, substantial changes are necessary. A better understanding is needed as to why people change their habits, how best to communicate with them, and what messages to communicate. Nutritionists can't do this job alone. Social scientists should have a part in designing and directing nutrition education efforts; psychologists and specialists in education should participate as well. Nutrition education can be effective only through an interdisciplinary approach.

BETTER CARE FOR CHILDREN

A special subject for nutrition education is how to feed children properly. The first chapter of The Nutrition Factor (Alan Berg) referred to malnutrition as one of the main causes of child mortality in developing countries.

MALNUTRITION IN HIGHLY DEVELOPED INDUSTRIALIZED COUNTRIES

Malnutrition as a public health problem is not confined to the developing countries. Overeating, which is a paradoxical form of malnutrition, can have almost as serious consequences as underfeeding; obesity and overweight have become a big problem among people whose food supplies are ample and whose energy requirements have been reduced because of a decrease in heavy work. Obesity is associated with various diseases, i.e., cardiovascular disorders and metabolic diseases such as diabetes mellitus, gout, etc.

The statistics of the Metropolitan Life Insurance Company show that for a man of 45 years of age, an increase of 10 to 12 kg above the standard weight reduces his life expectancy by 25 percent. This makes him likely to die at age 60 when he otherwise might have lived to age 80, had he not been obese. It should be emphasized that obesity is the most common nutritional disorder in the most industrialized countries. This means that even if there were no other nutritional problems in these countries, the training and knowledge of nutrition specialists would be justified.

The health authorities in the most industrialized countries have recognized the importance of nutrition education and have either created relevant institutions for training and education or have supported those institutions in societies who are concerned with nutrition education. Here I should like to give two examples:

- In Germany the German Society of Nutrition, which received ample funds from the Federal Government of Health, has two areas of main interest in the field of nutrition:

    - raising interest in the science of nutrition and spreading knowledge about the results of nutrition research

    - nutrition education of the consumer.

- A very potent organization has been reported from Great Britain. At the 1973 annual luncheon of the British Nutrition Foundation, Sir Keith Joseph, then Secretary of State, Department of Health and Social Security (DHSS) said: "I should like to propose that the Foundation and my Department should embark jointly on an exercise to examine the state of education in nutrition in this country and to consider ways and means by which it might be furthered and improved."

As a result of this proposal, a working party was set up early in 1974 to consider the problem. The group included the Officers of the Foundation, the Director-General of the Health Education Council (HEC), the Senior Medical Officer, Nutrition, and the Assistant Secretary of the DHSS, who was appointed chairman. There were three exploratory meetings in the first half of 1974.

Later in the year the Officers of the Foundation were approached by two senior members of Her Majesty's Inspectorate (HMI) who expressed a concern over the lack of effective nutrition education in schools. Following a preliminary discussion the Director-General and the Scientific Information Officer were invited to attend and present papers to a Department of Education and Science conference of HMI early in January. Discussions at this conference confirmed the conclusion that there was a lack of communication among all these bodies and much more might have been achieved if there had been coordination among these groups.

Because of this evident lack of coordination and communication, the Director-General of the HEC proposed that a meeting should be held early in 1975 to discuss problems with all the organizations identified as having an interest in nutrition education; community dieticians, health education officers, and representatives from the relevant ministries.

It is hoped that these two conferences and a national course for teachers in 1976 on nutrition education will be the beginning of more effective means of communicating modern knowledge from the specialists in nutrition science to other professional people.

SUMMARY

The whole field of training and education in nutrition is so broad, and has so many aspects and facets that I was able to touch on only some of them. If I was not able to describe all examples where nutrition is taught or should be taught, that does not mean that I felt they were of minor importance. My concern was to show by some examples the importance of teaching nutrition and to leave it to the members of this conference to draw conclusions for the respective fields and interests.

In preparing the manuscript, the following publications have been used:

1. Berg, Alan, The Nutritional Factor. Washington, D. C., The Brookings Institution, 1973.

2. Bosley, Betlyn, Nutritionists and dietitians in the seventies: trends in education. World Rev of Nutr and Diet 20:49-65, 1975.

3. FAO-WHO publication: Learning better nutrition, 1967.

4. Humanity and subsistence. Annales Nestle, 1961.

5. Working party on nutrition education. Brit Nutr Found Bull 14:May, 1975.

Reports from Missions and Meetings:

1. International Seminar on Education in Health and Nutrition, 1955.

2. FAO Nutrition Meeting for Europe, 1958.

3. Symposium on Education and Training in Nutrition in Europe, 1959.

4. Report of an FAO/WHO Mission, 1961.

5. Joint UNICEF/WHO Seminar on the Teaching of Nutrition in Mediterranean Countries, 1970.

FOOD AND HEALTH: CONSIDERATIONS OF THE PROTEIN METABOLISM WITH SPECIAL REFERENCE TO AMINO ACID REQUIREMENTS AND IMBALANCE

Hugo E. Aebi, M. D.

Professor of Biochemistry
Director, Medizinisch-Chemisches Institut der
Universität Bern
3000 Berne 9, Switzerland

Health versus hunger is the most drastic and also the most important contrast in today's world. It is no exaggeration to state that one of the main conditions for man's health is the availability of a well balanced nutrition in adequate quantity. There are three reasons to focus these considerations on proteins:

-- They represent - as indicated by the name given to this class by Muldern in 1838 - the body constituent of primary importance. This is true not only quantitatively, but also qualitatively, since all enzymes, the catalysts of living matter, belong to this class.

-- As a food component they have the unique property to be a source of amino acids required for protein synthesis as well as to serve as an energy donor, both aspects being closely related.

-- World wide shortage in supply is most pronounced in this class of foods. The following discussion on the relationship between food and health focuses on the production of more, inexpensive, edible proteins, since this is the most significant area to attack in the battle against hunger.

BASIC FACTS AND DEFINITIONS

Protein metabolism in normal man is characterized by more or less rapid turnover of proteins and amino acids in all organs and compartments of the body. The turnover rate varies over a wide range, the average half-life time varying between several months, e.g., collagen in skin, connective tissue, bones and blood vessels; a few days, e.g., intestinal epithelium; and a few hours only, e.g., enzymes and protein hormones. Although protein metabolism is a good example of how material can be economized by recycling, there is a continuous need for it in order to compensate for nitrogen losses. In the normal adult the average protein intake accounts for only a relatively small fraction of the daily protein turnover due to continuous synthesis and breakdown. With regard to the quantitative aspect it is assumed that the overall turnover rate is equivalent to a replacement of one to two percent of body protein each day; this corresponds to an average half-life time of total body protein of about 80 days. Since the overall protein content of adult man is 12 to 15 kilograms this would mean that the total body protein synthesis amounts to 200 to 300 grams per day. On the other hand, the obligatory nitrogen loss is 45 to 60 mg N (or equivalent to 0.3 to 0.4 g protein) per kg of body weight. This figure represents the sum of urinary and fecal nitrogen output, and of skin and other obligatory nitrogen losses. The data are based on observations made in healthy humans on a protein free diet for a period of at least 10 to 14 days (1-6). Consequently, the ratio between internal N-turnover and obligatory N-loss is of the order of ten to one.

The obligatory N-loss is a suitable reference - by definition - for the calculation of the minimal protein requirement; however, the composition of urinary and fecal N-products may vary considerably. If protein intake is very low, urea is no more the main endproduct of N-metabolism. Whereas on a strictly vegetarian low protein diet urinary-N is eliminated in almost equal amounts as urea, ammonia salts, hippuric acid, creatinine, uric acid, and other amino-N-compounds, a relatively large quantity of protein-N, probably of bacterial origin, is lost in feces (7). Recent studies made in malnourished children by means of 13 C and 15 N-labeled compounds, such as amino acids, urea, and ammonium salts, indicate that all of them enter the amino acid pool and that urea is utilized for microbial protein synthesis in the intestine (8, 9).

The most important figure used today as a guideline in defining an adequate nutrition is the "safe level of protein intake." This figure indicates the amount of protein considered necessary to meet the physiological needs and maintain the health of nearly all persons in a specified group (1). Since it depends on age, sex, and constitution it is subject to considerable individual variation. Furthermore, there is an interdependence between the level of protein intake and the intake of energy-bearing food (carbohydrate and fat). The specification of this figure according to age and sex, the assumption that the intake of energy carriers other than protein is not limiting, as well as a generous allowance of 20 to 30 percent contribute to the reduction of an unavoidable uncertainty. Such figures, notably those published in 1973 by the mixed Food and Agriculture Organization/World Health Organization (FAO/WHO) working group are a most valuable guide for the definition of an adequate nutrition and for the assessment of a given diet (1). The actual figures for the "safe level of protein intake," are slightly higher than earlier recommendations for the average protein requirement (10) (Figure 1).

AMINO ACID REQUIREMENTS AND BIOLOGICAL VALUE OF PROTEINS

The main aim of food science is to establish a balance between the daily requirements of the individual for all essential amino acids as well as total protein-nitrogen on one hand and the actual supply in food on the other. These requirements have to cover the needs for growth and maintenance. Three techniques have been extensively used in order to decide whether an amino acid is essential or not and how much of it is required under given conditions (11). Growth experiments in rats using diets containing unbalanced or incomplete amino acid mixtures or proteins give clearcut results, but caution is indicated in the application to man. Two other procedures, however, are suitable for studies in humans: the factorial method, which is based on estimates of the obligatory nitrogen losses and on

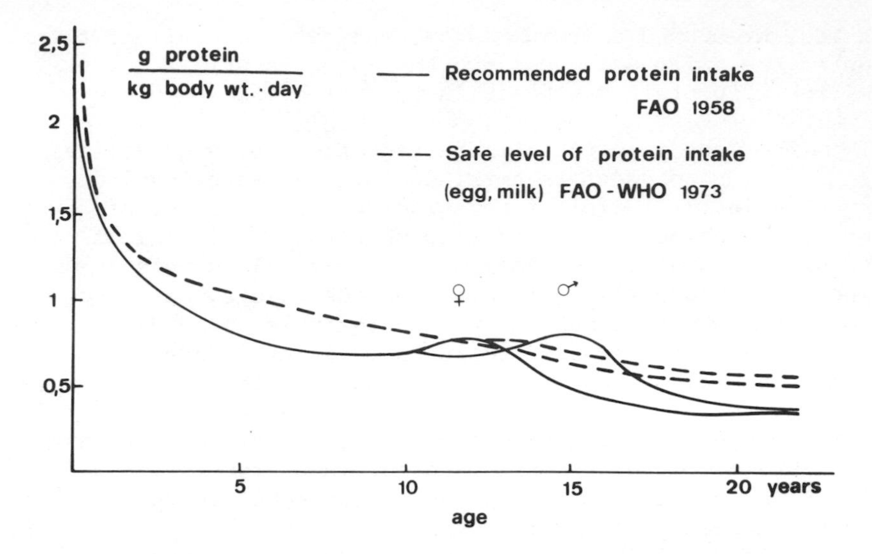

Figure 1 Official recommendations for the protein intake of infants, children and adults (1, 10).

the amounts of nitrogen needed for the formation of new tissue (adolescence, pregnancy, reconvalescence); and the measurement of the lowest protein intake needed to maintain nitrogen equilibrium in adults or satisfactory growth and N-retention in children.

In spite of the close interconnection of the quantitative and qualitative aspect of protein supply one can consider first the intake of an "ideal" protein, i.e., a protein the composition of which closely corresponds to the physiological needs. As an example, the safe level of intake of egg and milk protein has been taken in the report of the joint FAO/WHO expert committee (1). The data presented in Figure 1 (drawn according to the set of figures given in Table 23 of the FAO/WHO report) are based on measurements of the obligatory total nitrogen losses (factorial method), to which an additional 10 percent was added to allow for increased N-output related to minor infections and other sources of stress in daily life. Furthermore, 30 percent was added as an allowance for individual variation in order to reach the safe level of intake.

This allowance for individual variation is justified not only from a practical, but also from a scientific point of view. The concept of biochemical individuality is no longer an ill-defined term but can be based today on precise observations and experimental facts. The enzyme polymorphisms detected so far (20 loci out of 71 loci tested) permit the conclusion that heterozygosity for a variant or an anomaly is much more frequent than anticipated. The available data for 12 enzymes (15 loci) indicate that the probability for two individuals being genetically identical regarding their enzyme pattern is about one in six thousand (12). Extensive studies on large groups done at intervals offer good evidence that in nutrition, too, true differences among individuals (inter-individual variation) as distinct from the day-to-day variation (intra-individual variation) do exist (13, 14).

The quality of a protein in a given food can be tested either by analyzing its chemical composition, notably its content of essential amino acids, by balance studies in rat or man, or by using the growth rate of rats as an indicator. Two parameters are mainly used for this purpose:

-- The net protein utilization (NPU), which is defined as the percentage of ingested nitrogen retained in the body for growth, repletion, or maintenance. Usually this figure, which is the product of biological value of the protein and its digestibility, is correlated to the effect of a standard protein, such as casein (relative nitrogen utilization).

-- The protein efficiency ratio (PER), which represents the daily growth rate in grams of body weight related to protein consumption.

The significance as well as the variability of the protein efficiency ratio was recognized as early as 1909. Thomas defined the biological value of N-containing food as the number indicating how many parts of body-nitrogen can be formed out of 100 parts of food-nitrogen. One appreciates this view if one remembers that the complete series of amino acids of practical importance for nutrition has been known only since 1935, when Rose et al. succeeded in detecting threonine. Since then a large number of experimental data obtained from rats as well as observations made in humans have been accumulated. They permit one to conclude that for the adult, eight amino acids are essential, whereas for the infant, histidine has to be added.

The biological value of single proteins in food has been measured under a variety of conditions by different techniques. Biological methods as well as chemical analyses of the amino acid composition (amino acid score) permit the establishment of a scale of biological values of the major food proteins. Although there are minor differences they all lead to essentially the same result, irrespective of whether the amino acid score, the net protein utilization, or the protein efficiency ratio is taken as an indicator (1). Obviously the proteins at the top of this list contain these 8 [9] amino acids in almost exactly the proportion as they are needed by the body. At the bottom of the list are those proteins deficient in one or more essential amino acids. This is true in, e.g., gliadines (deficient in lysine) and proteins in leguminoses (deficient in methionine). An extended list of biological value data permitting a comparison of seven different indices and ratios proposed for rating the protein quality, has recently been published by Mauron (15). Whereas the amino acid score gives a clear-cut answer about the sequence of the limiting amino acids in a given protein, the NPU and PER-data indicate to what extent this protein can be utilized by the growing or the adult organism, fed as the only source of amino acids in a diet. This aspect is of practical relevance only when a diet is monotonous in its composition. Unfortunately this is true for large regions. In most cases, however, the average diet contains several proteins of major importance and, therefore, the amino acid pattern of the resulting protein combination becomes essential.

THE UTILIZATION OF PROTEIN COMBINATIONS IN MAN

For the assessment of the protein supply in a well balanced diet all contributing proteins have to be considered. N-balance experiments are of crucial importance in humans, even if they are tedious and time consuming and their reliability is sometimes questioned. They represent the only approach which can have a reliable and realistic answer. The studies made at the Max-Planck-Institute for Nutrition Research in Dortmund, GFR, by Kofranyi may serve as an example. With the help of a number of devoted volunteers, he has performed a remarkable series of N-balance studies. By analyzing protein mixtures he was able to show that, in almost all instances, the two components act in such a way that the combination gave better results. This increase in utilization can be observed not only in protein-pairs of high and low biological value, such as milk-wheat or egg-maize, but also among proteins of low biological value, such as maize and bean-protein (16, 17, 18). Provided that these findings are valid and can be extrapolated without restriction, one may deduce that less protein is required to maintain the nitrogen equilibrium in the body, if it is consumed in a well balanced mixture.

By systematic variation of the ratio of the components, a combination was found requiring the smallest quantity of protein necessary for the maintenance of nitrogen equilibrium. The smaller the quantity required for maintenance of nitrogen equilibrium, the better protein utilization. A typical example is the combination of milk and wheat. Here the required quantity of protein is smallest when a mixture of 75 percent milk protein and 25 percent wheat protein is consumed. The lowest protein intake, i.e., the best utilization, however, is observed after feeding a mixture of 36 percent egg protein and 64 percent potato protein.

What conclusion can be drawn from such studies? A considerably better protein utilization can be achieved in adult man by combining proteins in optimal proportion, particularly by "diluting" proteins of high biological value with a suitable partner of (theoretically) low biological value. Since most earlier investigations of that kind have been performed in rats, it must be noted that there is a species difference, inasmuch as this effect of economy in combining proteins seems to be considerably smaller in the rat than in man. Consequently, the content of essential amino acids in animal proteins, such as meat, egg, and milk, is probably too high to be fully utilized by a healthy adult man. Recent clinical studies point in the same direction (19). Further support for the idea of combining proteins rich in essential amino acids with proteins of an incomplete or insufficient pattern of essential amino acids comes from the analysis of the ET-ratio (essential amino acids related to total protein content) (1). The ET-ratio in animal proteins exceeds the recommended level, at least for the adult human. Furthermore, there is no reason to disqualify plant proteins just because their biological value is lower than that of animal proteins. Both of them are valuable components in man's food provided they are consumed in a reasonable proportion.

AMINO ACID IMBALANCE AND AMINO ACID FORTIFICATION

Considerable effort is made today to optimize protein utilization, not only in human nutrition, but also in modern husbandry. In order to get maximal meat production some sort of "amino acid mathematics" has been developed. Most of these studies, whether performed in agricultural or biomedical research, were based on the assumption that each essential amino acid represents an entity in itself and is handled in metabolism independently. More recent experiments have shown, however, that a number of amino acids interact with each other, an interdependence which is in certain cases of practical relevance.

The first evidence that single amino acids may interact or even compete with each other in their growth promoting effect was obtained by Elvehjem *et al.* and Harper (20, 21). They observed that the addition of gelatine or threonine to a 9 percent casein-diet, lacking niacine, resulted in a severe depression of growth rates. This effect could be reversed by a supplement of niacine or tryptophane. Similar adverse effects of amino acid supplements were subsequently observed in experiments on a variety of dietary proteins with mice, rats, dogs, pigs, and chicks. There are several types of amino acid imbalance known today.

A depression of the growth rate can be induced either by the addition of a nonbalanced protein or amino acid mixture, or by adding an essential amino acid (e.g., threonine) in significant amounts to an otherwise balanced diet. The organism seems to be particularly sensitive to the balance between the first and second limiting amino acid, notably if a low protein diet is fed. Amino acid antagonism is another type of imbalance, which is probably due to a competitive action among structurally related partners, such as leucine, isoleucine, and valine (21, 22). Part of this adverse effect on the growth rate is due to a sensible reduction of the food intake, which normally is rather high in young rats. It is uncertain to what extent similar consequences of inadequate food combinations may occur in man (23). In many countries humans prefer boiled, fried, or otherwise processed food, so another discrepancy between amino acid composition and biological value may occur. In the Maillard reaction between the ε-NH_2-group and carbohydrate partner molecules, lysine becomes partly blocked and, therefore, is no longer available (24, 25). There is an analogous situation for methionine in plant material (26).

There are two ways of improving the amino acid supply in a deficient diet. First, protein combinations, which permit the best possible utilization of their amino acids, should be chosen. If this is not possible or feasible there remains the possibility of amino acid fortification, i.e., the addition of pure amino acids (27, 28). This procedure is indicated when the first or second limiting amino acid is deficient or missing in an otherwise acceptable food. Although first experiences with lysine (modern bread in India) or lysine + methionine (broth cubes in Nigeria) are encouraging, the long range beneficial effect of such additions has to be carefully studied under a variety of different conditions before a generalized distribution can be recommended. The effects of the addition of single amino acids to a diet are not predictable; therefore, they have to be verified by experiment. The addition of lysine and methionine raises the protein efficiency ratio of a rice/bean diet almost to that of casein (28). However, the net gain in protein utilization is relatively modest. Despite this limitation it must be admitted that a regional deficit in lysine can probably best be overcome by a specific fortification of basic food with lysine or lysine-enriched products. On the other hand, the world-wide problem of an inadequate supply of essential amino acids could best be solved by more efficiently combining proteins of complementary composition, the difficulty being, however, the acceptability of certain proteins. Fortification is an acceptable and desirable procedure as long as it fits the general framework of actions aimed at a harmonization of the amino acid supply.

MECHANISMS REGULATING PROTEIN METABOLISM

Although Cannon's principle of homeostasis has been known since 1932 and the molecular mechanisms of amino acid metabolism and protein synthesis are known in some detail, there is still little insight into how protein metabolism is coordinated (29). The mutual adaptation of the rates of protein synthesis and breakdown is obviously one of the clues to how the body of adult men manages to maintain a perfect nitrogen balance (30). Probably the regulation of enzymatic activities plays an essential role in accomplishing this task. Whereas allosteric inhibition brings about instantaneous alterations of both enzyme-substrate affinity and specific activity, the adjustment of the level of enzyme concentration is a slower, but most efficient regulatory mechanism. An alteration of enzyme concentration may be the consequence of a change in the rate of enzyme synthesis or of enzyme degradation (30). Since the synthesis (and probably also the breakdown) of all proteins is under genetic control, it is evident that genetic regulatory mechanisms, such as induction and repression, play an essential role in continuously adapting the metabolizing capacity of the organism to the actual environmental conditions. A good example, showing how a whole sequence or cycle of enzymes is regulated in a concerted manner, is the group of enzymes involved in urea synthesis (31, 32, 33). In his classical study Schimke showed in 1962 that the level of all five urea cycle enzymes in rat liver is strictly proportional to the daily protein consumption. It is tempting to conclude, from observations of this type, that close relationships do exist between the level of protein intake and all enzymes involved in its metabolism. In more recent studies Das and Waterlow have shown that drastic adaptive changes in arginase activity are completed within 30 hours (34). Since related enzymes, such as glutamic dehydrogenase, are not affected at all, it is obvious that these rapidly occurring alterations in enzyme concentration are the result of specific regulatory signals. Further studies of Schimke have shown that an increase in tryptophane pyrrolase in rat liver can be accomplished by corticoid hormones as well as by feeding a diet high in tryptophane. The former acts by increasing the rate of synthesis, the latter by blocking the degradation (30).

Considering the multitude of enzymic reactions involved in adapting the pattern of amino acids absorbed in the intestine to the actual needs of the body, efficient and versatile regulating systems must be operating. A well investigated example is the regulation of a couple of enzymes involved in serine synthesis and degradation. Mauron _et al_. (1973) investigated the effect of different dietary proteins on the level of serine dehydratase, E. C. 4.2.1.13 (involved in the serine-pyruvate pathway, a step in the conversion of protein to carbohydrate) and 3-phosphoglycerate dehydrogenase, E.C. 1.1.1.95 (involved in the pyruvate-serine pathway, thus participating in the conversion of carbohydrate to protein). When being fed a high protein diet, rats became adapted by induction of serine dehydratase and repression of phosphoglycerate dehydrogenase (and vice versa when fed a low protein diet). This is a logical regulatory mechanism, since, with a high protein diet, the excess amino acids are converted to carbohydrate in order to produce energy; with a low protein diet carbohydrate must be converted to amino acids as much as possible. Both enzymes react in an opposite, but perfectly symmetrical way and can be used as markers of the state of protein metabolism. Quantitative differences seen when feeding casein, gluten, and single cell protein prompted the authors to extend this study to artificial amino acid mixtures in order to explain the observed discrepancies. They showed that not all ten amino acids essential for growth of the rat were active in regulating the level of these two enzymes. Only methionine, tryptophane, threonine, and valine were active in inducing serine dehydratase and repressing phosphoglycerate dehydrogenase (35).

These and other similar observations suggest a distinction between essential amino acids and regulatory amino acids. The former group contains all amino acids which cannot be synthesized in the body; the latter group, only four amino acids, is responsible for the enzymatic adaptation of the organism to variations in the protein supply. Consequently, the concentration of regulatory amino acids in a given diet mainly initiates the adaptive response of the body, and the total concentration of all essential amino acids controls total nitrogen retention. Although the validity and the implications of this working hypothesis remain to be established, these observations confirm, beyond any doubt, that specific regulatory mechanisms within the amino acid interconversion system exist to enable the organism to make the best use of the proteins in food. The fact that lysine - although often limiting the biological value of protein - is not a regulatory amino acid, explains the observation that casein and gluten, in spite of the great difference in nutritive value, behave similarly in this respect (35).

Activity and concentration of a number of enzymes in rat liver are affected by quantity and quality of food and consequently also by the rhythm of food intake (36). Van Potter has analyzed 12 enzymes in this respect and has shown that there are considerable rhythms regarding the feeding schedule, but that there is also a circadian rhythm. By adapting rats to a controlled feeding schedule (i.e., 16 of 40 hours fasting period and 8 hours feeding period), thus considerably decreasing the degree of data variation, he could demonstrate that clearcut tidal changes in enzyme pattern do occur. Oscillations of that type are seen, e.g., in tyrosine transaminase, serine dehydratase, and glucose-6-phosphate dehydrogenase (36). The existence of a circadian rhythm in concentration of plasma amino acids, as well as their excretion in urine, has also been demonstrated in man. This periodicity is characterized by peak values between noon and 8 p.m. and low values in the early morning (4 to 8 a.m.); however, there are differences among the single amino acids (37-40). Furthermore, the response of plasma amino acid levels after a test meal high in protein (ca. 100 grams) depends considerably on the time of the day and, therefore, the phase of the rhythm. The ingestion of this relatively large protein load at 8 a.m., i.e., in the rising phase of the rhythm, resulted in a small, but significant additional increase; the same test meal eaten at 8 p.m. did not affect the plasma amino acid level in its declining phase (37). It would be of interest to know whether this difference correlates to man's readiness to consume a meal rich in protein as a function of day time activity. In control studies Feigin (38) has shown that there is a lack of direct relationship between dietary protein intake and blood amino acid values and that normal periodicity is maintained even during starvation for several days. On the other hand, amino acid rhythmicity is easily altered by the onset of an acute illness or a change in the sleep-awake pattern. Since the excretion of 17 hydroxycorticosteroids closely follows the amino acid rhythm, it may be concluded that the amino acid level in plasma - as in a number of other biochemical parameters - reflects the periodicity of mental and physical activity, as well as the metabolic processes in the body (40). As stated in the ARC-MRC Report (2) "the problem of the coordinated regulation of turnover rates is the one which presents the greatest challenge for future work in the whole field of protein metabolism and nutrition; . . . the challenge can only be met by integrated studies in the whole animal and at the cellular and subcellular levels."

ASSESSMENT PROBLEMS

Guidelines of experts and theories on biochemical mechanisms are senseless unless an energetic effort is made to improve the nutritional situation, whenever the safe level of protein intake is not reached and the risk of serious metabolic consequences is imminent. Since it is neither possible nor feasible to establish a nitrogen balance in all such instances, simpler, though indirect, methods have to be used in order to find out whether, in an individual or in a population group, protein intake is sufficient or not. Today, for practical reasons, anthropological data are mainly used for this purpose. However, since exact and reliable anamnestic data on food intake are lacking in many cases, there is a definite need for biochemical parameters. The search for quicker and simpler diagnostic tools as an indicator for adequate or insufficient protein supply is of great practical interest. This diagnostic help from the laboratory would be particularly appreciated in marginal or subclinical cases of protein calorie malnutrition. The parameter which is most intensely investigated and the study which has so far given much valuable information, is the concentration of the free amino acids in plasma (urine or eventually feces), either as the complete plasma aminogram (39, 41) or as the amino acid ratio according to Whitehead (42, 43).

One of the main obstacles - besides large biological variation due to individual differences and circadian rhythms - is the limited accessibility of suitable material. In order to avoid liver or muscle biopsy, the analysis of hair follicles, hair, nails, mucosa clippings, etc., has been attempted, so far without much success. The study of enzymes and metabolites in blood elements, therefore, seems to be most promising. Due to the rapid turnover rate and their rich enzyme pattern, leucocytes, in spite of methodological difficulties, are likely to be more suitable than erythrocytes, at least in this respect (44, 45). In the report, "Food and Nutrition Research," recently published by the Agricultural Research Council (ARC) and the Medical Research Council, (MRC) it is stated: "From the practical point of view, if enzymes which show clearcut adaptive changes in response to diet could be measured in blood or tissues, this would constitute an important diagnostic tool" (2). The observation recently made in this laboratory that, unlike glucose-6-phosphate dehydrogenase activity in rat liver, the leucocyte enzyme level (related to total-N) is practically independent of the level of protein intake, is not very encouraging. Whether leucocyte catalase is a suitable indicator, as seen in a study on the white cell enzyme pattern of patients on a zero-caloric diet, remains to be seen. More data obtained by applying the method recommended by the IUNS working group would be highly desirable. Even more important than a plea for more data is the recommendation given recently in the ARC-MRC Report mentioned above, that better and standardized methods be used, which "compared with present crude animal procedures, are more sensitive and refined, and which are subject to fewer uncontrolled factors" (2).

OUTLOOK

In a report delivered to the International Organization for the Study of Human Development, it is appropriate to confront the wealth of theoretical knowledge with the actual practical situation and to extrapolate the conclusions from the single individual to the entire population. The problems related to food and health cannot be solved without science, but they overlap by far the scope of basic and applied research.

In concluding, a few crucial questions which are subject to controversy, shall be briefly asked.

How much protein is needed?

Although excellent guidelines, based on a large number of measurements and thoughtful considerations, have been given by competent committees, the controversy around this figure is not likely to stop. Official nutritional standards, such as the "safe levels of protein intake," are a valuable tool to assess the adequacy of a particular diet and of national food supplies. They include a considerable allowance in order to account for interindividual variation, but they are intended as a lower warning limit. Such standards represent an important and realistic aim for developing countries. In industrialized countries this "safe level" is by far overshot by most individuals; they usually consume the two or three-fold quantity compared to the safe level. Their food, as an average, contains 11 to 13 percent protein (in terms of calories), whereas the same figure for mother's milk - despite the high demand for protein of the infant - is only 8 percent. The requirement for protein of the adult is often overestimated. A review on the controversial aspect of the terms _minimal_ and _optimal_ level of protein intake has recently been given by R. Bircher, who, for many years, has been a devoted fighter against too generous allowances made in defining an optimal level of protein intake (46). However, what may be true for a disciplined consumer cannot be applied to the whole population. On the other hand, the physiologically high demand for protein of the infant and the child cannot be satisfied in many instances due to lack of suitable food. Therefore, whenever "safe levels" are in danger, clearcut priority has to be given to the child.

Does more protein in food necessarily mean better health?

Irrespective of the absolute level there is a world-wide trend to eat more protein, especially animal protein. In Switzerland, e.g., the average meat consumption has risen in the past 25 years from 45 to 75 kg/person/year, the cost of which amounted in 1974 to 25% of the total expense for food (47, 48). Comparative studies between various countries and among different social groups have shown that there is an almost proportional correlation between the gross national product (or average income) and protein consumption. It is imperative to improve the economic situation and, thus, to increase the protein supply in food for all those humans, who have not yet reached the required minimal level! This does not need to be discussed again. In industrialized countries this change in eating habits - more meat, less cereals, bread and potatoes - is connected with the altered living style, with its stress and hurry, as well as the role of meat as an ergotropic stimulant or just a status symbol.

It is difficult to decide whether this trend to more protein and to animal protein is linked with acceleration and longevity. There is always a temptation to extrapolate the result of experiments with rats to the human organism (e.g., the stimulating action of a diet high in protein on the growth rate). Doing so one has to remember that the purpose of human nutrition is quite different: Man expects a long life with a minimum of disease and a maximum of intellectual and/or physical capacity throughout his life. Obviously these objectives have little - or even nothing - to do with growth rate and body size. Likewise the ideas that maximal enzyme activity or a high degree of N-retention represent a state most beneficial to the organism and, therefore, should serve as guidelines for realizing optimal living conditions, must be questioned for rat and even more so for man. Experimental evidence is accumulating that in rats a negative correlation between maximum growth rate and longevity does exist (McCay, Ross in (49)). More recently Mauron has found that the change in activity level of some enzymes (e.g., xanthinoxidase, D-amino-acid oxidase) with age is similar to their response to increased dietary protein quantity or quality (50). Clearly, these observations cannot be extrapolated to man but show that the "bigger and better concept" cannot be applied to protein intake of man.

Is there a unifying concept?

Among these conflicting tendencies a reasonable compromise must be found. Looking at proteins, notably animal protein, two driving forces are opposing each other. They are - if a gross oversimplification is tolerated - man's trend to joy and pleasure on one hand, and his lack of motivation to avoid waste on the other.

Food in (animal) protein content provides a threefold pleasure by its pleasant look, odor, and taste; by its high sensorial value, accompanied by easy digestibility and a high saturation value; and its well known stimulating effect, mediated by an increased catecholamin liberation, in intellectual as well as physical activity.

However, intake of food high in animal protein content also represents a threefold waste - biochemically, financially, and economically:

-- Gaining energy from amino acids requires a number of detoxifying reactions, such as deamination, desulfuration, and urea synthesis. Total oxygen consumption is increased due to the specific dynamic action, probably because oxidation of amino acids is thermodynamically a less efficient process.

-- All animal proteins are relatively expensive; only a privileged minority can afford an excess protein intake.

-- Considerably larger areas of land and more labor are required to produce the same quantity of animal protein, compared with plant proteins (4, 5, 51). If a real priority is given to achieve maximal food (and protein) production, this discrepancy cannot be overlooked any longer.

In being realistic one has to admit that these antagonisims will probably never be overcome completely. However, much can be done to make them compatible. For the scientist in basic research there are mainly three problems deserving a special effort:

-- Searching for new protein sources and for better protein combinations.

-- Gaining more insight into the regulatory mechanisms involved in optimal utilization of the amino acids available in food proteins.

-- A better understanding of the motives involved in food selection and acceptance at the psychological level.

It is possible that more insight may contribute to finding a reasonable compromise between man's driving forces aimed at satisfaction and pleasure on one side and his willingness to reduce his own demands and to cooperate on the other. The answer to the question of how food and health are related is simple: For a privileged minority, better health means *less* food; for the underprivileged majority of the world's population better health means *more* food, above all *more* protein!

REFERENCES

1. Report of joint FAO/WHO Committee: Energy and Protein Requirements. Geneva, WHO Tech Rep Ser, No. 522, 1973.

2. Report of ARC/MRC Committee: Food and Nutrition Research. London, Her Majesty's Stationery Office, 1974.

3. Groupe de la Nutrition, OMS, Genève: Examen et utilisation a des fins nutritionelles des indicateurs actuellement disponible dans le domaine de la santé. PAG Bull 4:1-5, 1974.

4. FAO, La situation mondiale de l'alimentation et de l'agriculture 1974. Rome, 1975.

5. FAO, Examen de la situation alimentair mondiale - présente et future. Alimentation et Nutrition 1:7-41, 1975.

6. Scrimshaw NS, Hussein MA, Murray E, Rand WM, Young, VR: Protein requirements of man: variations in obligatory urinary and fecal nitrogen losses in young men. J Nutr 102:1595-1604, 1972.

7. Oomen, HA: Distribution of N and composition of N-compounds in food, urine and feces in habitual consumers of sweet potato and taro. Nutr Metabol 14:65-82, 1972.

8. Read WWC, McLaren DS, Tchalian M: Urinary excretion of nitrogen from ^{15}N-labelled amino acids in the malnourished and recovered child. I.glycine and lysine. Clin Sci 40:375-380, 1971.

9. Read WWC, McLaren DS, Tchalian M: ^{15}N studies of endogenous fecal nitrogen in infants. Gut 15:29-33, 1974.

10. FAO: Rapport du Comité de la FAO, Besoin en protéines. Rome, 1958.

11. Irwin MI, Hegsted DM: A conspectus of research on amino acid requirements of man. J Nutr 101: 541-566, 1971.

12. Harris H: The principles of human biochemical genetics, 2nd edn. Amsterdam, North Holland/ Elsevier, Oxford, 1975, p. 296.

13. Calloway DH, Margen S: Variation in endogenous nitrogen excretion and dietary nitrogen utilization as determinants of human protein requirement. J Nutr 101:205-216, 1971.

14. Beaton GH, Swiss LD: Evaluation of the nutritional quality of food supplies: prediction of desirable or safe protein: calorie ratios. Amer J Clin Nutr 27:485-504, 1974.

15. Mauron J: The analysis of food proteins, amino acid composition and nutritive value. In, Porter, Rolls (eds): Proteins in Human Nutrition. New York, Academic Press, 1973, pp. 139-154.

16. Kofranyi E, Jekat F: Die Bestimmung der biologischen Wertigkeit von Nahrungs-Proteinen, XIV; die Mischung von Rindfleisch und Gelatine. Hoppe Seyler's Z. Physiol Chem 350:1405-1409, 1969.

17. Kofranyi E, Jekat F, Müller-Wecker H: The minimum protein requirement of humans, tested with mixtures of whole egg plus potato and maize plus beans. Hoppe Seyler's Z. Physiol Chem 351:1485-1493, 1970.

18. Müller-Wecker H, Kofranyi E: Einzeller als zusatzliche Nahrungsquelle. Hoppe Seyler's Physiol Chem 354:1034-1042, 1973.

19. Kopple JD, Swendseid ME: Nitrogen balance and plasma amino acid levels in uremic patients fed on essential amino acid diet. Amer J Clin Nutr 27:806-812, 1974.

20. Elvehjem CA, Krehl WA: Dietary interrelationships and imbalance in nutrition. Borden's Rev Nutr Res 16:69, 1955.

21. Harper AE: Amino acid balance and imbalance I; dietary level of protein and amino acid imbalance. J Nutr 68:405-418, 1959.

22. Mauron J: Amino acid imbalance and its bearing on the fortification of food. Bibl Nutritio et Dieta 11: 57-76, 1969.

23. Oezalp I, Young VR, Nagshandhuri J, Tontisirin K, Scrimshaw NS: Plasma amino acid response in young men given diets devoid of single essential amino acids. J Nutr 102:1147-1158, 1972.

24. Mauron J, Mottu F: Relationship between in vitro lysi lysine availability and in vivo protein evaluation in milk powders. Arch Biochem Biophys 77:312-327, 1958.

25. Finot PA, Mauron J: Le bloquage de la lysine par la réaction de Maillard I et II. Helv Chim Acta 52: 1488-1495, 1969; idem 55:1153-1164, 1972.

26. Devenyi T, Bati J, Hallström B, Tragardh C, Kralovansky PU, Matrai T: Determination of "available" methionine in plant materials. Acta Biochim Biophys Acad Sci Hung 9:395-398, 1974.

27. Mauron J: Amino acid imbalance and its bearing on the fornification of food. Bibl Nutritio et Dieta 11:57-76, 1969.

28. ___________: Protein enriched foods: facts and illusions. Proc 9th Int Congr Nutrition, Mexico, 1972, 3:231-245, Basel, Karger, 1975.

29. von Muralt A: Protein-calorie malnutrition viewed as a challenge for homeostasis. Protein-Calorie Malnutrition, a Nestlé Foundation Symposium. Berlin-Heidelberg-New York, Springer Verlag, 1969.

30. Munro HN (ed): Mammalian Protein Metabolism, V. 4. New York, Academic Press, 1970.

31. Schimke RT: Adaptive characteristics of urea cycle enzymes in the rat. J Biol Chem 237:459-468, 1962.

32. Aebi H: Coordinated changes in enzymes of the ornithine cycle and response to dietary conditions. In The Urea Cycle. New York, J. Wiley & Sons Publishers, 1975.

33. Aebi H: Enzymes and nutrition. Protein-Calorie Malnutrition, a Nestlé Foundation Symposium. Berlin-Heidelberg-New York, Springer Verlag, 1969.

34. Das TK, Waterlow JC: The rate of adaptation of urea cycle enzymes, aminotransferases and glutamic dehydrogenase to changes in dietary protein intake Brit J Nutr 32:353-373, 1974.

35. Mauron J, Mottu F, Spohr G: Reciprocal induction and repression of serine dehydratase and phosphoglycerate dehydrogenase by proteins and dietary essential amino acids in rat liver. Europ J Biochem 32:331-342, 1973.

36. Potter VR, Baril EF, Watanabe M, Whittle ED: Systematic oscillations in metabolic functions in liver from rats adapted to controlled feeding schedules. Fed Proc 27:1238-1245, 1968.

37. Feigin RD, Klainer AS, Beisel WR: Factors affecting circadian periodicity of blood amino acids in man. Metabolism 17:764-775, 1968.

38. Feigin RD, Beisel WR, Wannemacher RW: Rhythmicity of plasma amino acids and relation to dietary intake. Amer J Clin Nutr 24:329-341, 1971.

39. Hussein MA, Young VR, Murray E, Scrimshaw NS: Daily fluctuation of plasma amino acid levels in adult man: effect of dietary tryptophane intake and distribution of meals. J Nutr 101:61-70, 1971.

40. Tewksbury DA, Lohrenz FN: Circadian rhythm of human urinary amino acid excretion in fed and fasted states. Metabolism 19:363-371, 1970.

41. Snyderman SE, Holt, LE, Norton PM, Roitman E, Phansalker SV: The plasma amino/gram; I, influence of the level of protein intake and a comparison of whole protein and amino acid diets. Pediat Res 2:131-144, 1968.

42. Whitehead RG, Dean RFA: Serum amino acids in kwashiorkor I; relationship to clinical condition. Amer J Clin Nutr 14:313-319, 1964.

43. Arroyave G, Bowering J: Plasma free amino acids as an index of protein nutrition. Arch Lat Amer Nutr 18:341-361, 1968.

44. Yoshida T, Metcoff J, Frenk S, dela Pena C: Intermediary metabolites and adenine nucleotides in leukocytes of children with PCM. Nature 214: 525, 1967.

45. Munro HN: Leukocytes and fetal malnutrition. Pediatrics 51:926-928, 1973.

46. Bircher R: Zue Eiweissfrage Dieita-Beilage ze Erfahrungsheilkunde. Heidelberg, Verlag Haug, 1972.

47. Aebi H: Unsere Ernährung im Spiegel des Gesellschaftlichen Wandels. Heft 27 der Schriftenreihe der Schweiz. Vereinigung für Ernährung. Bern, Verlag der Schweiz, 1974.

48. Aebi H, Trautner K: Entwicklungstendenzen in der ernährungeweiss. Bibl Nutr Diet 21:11-23, 1975.

49. Ross HM: Length of life and nutrition in the rat. J Nutr 75:197-210, 1961.

50. Mauron J: The value of measuring enzyme activities in assessing the adequacy of a protein diet. Kuhnan J (ed): Proc Int Congr Nutr Food, Hamburg, 4:367-381. Viehweg & Sohn, Braunschweig; Oxford, Pergamon Press, 1967.

51. Abott JC: The efficient use of world protein supplies. PAG Bull 3:25-35, 1973.

FOOD AND GENETIC DEVELOPMENT

Jean Frézal, M. D.

Professor
Hôpital des Enfants Malades
149 rue de Sèvres
Paris, France

The relationship between food and genetic development raises questions which can conveniently be arranged under three headings:

-- Do foods, food additives, and food contaminants induce changes in the genetic information? That is, do they cause mutations?

-- Can food habits cause short-term or long-term modifications of the gene pool? That is, are they selective factors?

-- Do ingested nutrients influence the expression of the genetic program or interact with it? This question is, in some respects, related to the problem of genetic regulation.

In this paper we shall emphasize these three topics - the first two rather briefly, the last one more thoroughly.

INDUCED MUTAGENESIS

There is some experimental evidence that such substances as preservatives, artificial sweeteners, dyes and stimulants (like caffeine) used in food preparation and food technology are potent mutagens. Contaminating chemicals, such as the pesticides utilized in agriculture, and pollutants from various sources, may also be mutagenic (1).

However, one must be very cautious about drawing inferences from experimental work and applying them to human beings. This need for caution is well illustrated by the caffeine story (2): Caffeine is known to be mutagenic in bacteria, in plants, and in some Drosophila lines. It has also been demonstrated to induce chromosomal breakage when added to human cell cultures in vitro. The latter evidence has led certain authors to claim that caffeine is one of the most dangerous mutagenic agents for man - a statement not strongly supported by the data. In fact, surveys of children born to parents who were heavy coffee drinkers have thus far yielded no evidence of mutation.

The studies with caffeine illustrate the difficulties involved in determining whether a known or suspected mutagen present in food can significantly alter the genetic load of the human population which consumes that food. First, opportunities for detection of induced mutagenesis are limited to those genetic traits for which the spontaneous mutation rate is already known. Scanty information in this area, coupled with unresolved questions of methodology, places severe restrictions on the amount of knowledge to be gained from surveys of human populations.

Second, there is a dearth of information concerning the relationship between ingestion of a suspected mutagen and events at the cellular level. In vitro studies have shown that both the number of changes and the kind of changes (punctual or chromosomal) induced by a mutagen are determined by the concentration of the agent in the target cells, by the timing of administration relative to the cell cycle, and by the efficiency of cellular repair mechanisms. Too little is known about the role of these factors in vivo.

Finally, it must be remembered that the relationship between mutation and phenotypic change is not linear, except in the case of autosomal dominants.

Considering all of these facts, it seems unlikely that allegedly mutagenic substances present in food will be found to have any noticeable influence on the genetic load, at least for a very long time.

EVOLUTION AND SELECTION

The question of selection is quite distinct from the question of mutagenesis. Even if components of foods are unable to induce mutations at a significant rate, it is still possible that nutrition, foods, and food habits could act as selective factors, favoring some genes and eliminating others.

Intuitively, one may suspect that nutrition, and particularly starvation, must have been selective forces as important as infectious diseases in the history of humanity. It has been suggested that some polymorphisms, which are now apparently neutral, may have arisen in the past in response to selective action of either nutritional factors or infection. Clearly, this hypothesis cannot be tested directly, so it is necessary to use inductive reasoning to discover which genes can have been so affected.

No doubt the best and most illuminating evidence about the evolutionary influence of food habits and food supply is given by the lactase "saga" (4).

It seems likely that the loss of the capacity to synthesize substances such as vitamins and essential amino acids is the consequence of selection under two conditions: mutation and sufficient exogenous supply. It is probably advantageous for an organism not to manufacture products which are abundantly provided by the environment.

Adaptation of milk composition to the suckling infant or animal provides another example which may be accounted for by selection. The mechanisms of fat digestion and absorption are known to differ completely in man and ruminants. In man, the structure of ingested triglycerides is partially preserved. Therefore, the distribution of fatty acid residues at the different positions of the glycerol molecule has a significant effect on the rate at which a triglyceride can be solubilized and resynthesized. Thus it is understandable that natural fats are more readily absorbed in man when they contain less stearic acid (always found at the one or three position) and more palmitic acid (always at the two position). In ruminants, on the other hand, distribution of fatty acid residues on the glycerol molecule is of no significance, since triglycerides are completely hydrolyzed before absorption (5).

It seems reasonable to suppose that certain widespread diseases, such as atherosclerosis and diabetes mellitus, could have arisen from insufficient adaptation of the human genome to recent extension of the life span and overfeeding. This hypothesis seems, to this writer, more attractive than the alternative one relating the actual prevalence of these diseases to a recent relaxation of selection.

FOOD AND EXPRESSION OF GENES

That nutrition and foods may influence the expression of the genetic program is demonstrated by the dietetic management of such congenital errors of metabolism as phenylketonuria (PKU) and galactosemia.

In a study of the physical and mental development of 60 galactosemic children who had been maintained on a low-galactose diet, Komrower and Lee (6) found that the physical health of the group was good but that mental development fell below that of the average population, with a significantly higher incidence of psychological disturbances. According to these authors, this finding, taken together with the fact that the level of galactose-1-phosphate is elevated in the cord blood of galactosemic newborns, suggests an intrauterine disturbance of development. Visual-perceptual disturbances and/or learning difficulties may also result from such an intrauterine metabolic imbalance (7).

Evidence of this sort leads one to suspect that an inborn error of metabolism influences the expression of an entire array of genes. In order to determine whether this is the case, it is necessary to know more about the normal sequence of genetic events in prenatal development. Specifically, it is necessary to know the timing of expression and repression of key genes, to know whether embryonic enzymes and their postnatal counterparts are identical (i.e., whether they are coded for by the same locus), and to know whether embryonic and placental enzymes are identical.

A closely related problem concerns the manner in which an inborn error of metabolism in a pregnant woman affects the fetus. The two extreme possibilities are represented by phenylketonuria, in which an affected mother usually bears a mentally retarded child; and galactosemia, in which untreated, affected mothers bear normal children (8).

In this context, it is worthwhile to suggest that placental genetic defects must exist, and may well play some role in both spontaneous abortion and developmental disturbances. The well-known polymorphism of the alkaline phosphatase of the placenta hints at this possibility, although in this particular example, the large number of variants (indeed, the largest for any locus but the antigenic ones) does not suggest any differential selective value (9).

Finally, in addition to total or near total loss of enzymic activity, a number of mutations may be responsible for partial deficiencies which could be harmful only in certain circumstances, as suggested by the intermittent form of hyperlysinemia. On *a priori* grounds, one might expect a greater susceptibility of these mutants to environmental influence, and, therefore, a more irregular distribution within human populations. This is a new aspect of genetic variation, as yet barely unveiled, which in this author's opinion may be of great significance for human development.

The problem of the influence of minor genetic variations is linked to the difficult question of multifactorial traits. According to a widely accepted hypothesis, certain congenital malformations, along with such common diseases as diabetes and the hypercholesterolemias, and quantitative traits like stature and intelligence, may be polygenic in nature. That is, they may be determined by multiple genes which interact with each other, and/or the environment, to produce a threshhold effect. If this is correct, it seems likely that relatively minor deficiencies or imbalances in some components of the diet may be of significance in the causation of developmental defects of the brain.

The problem of determining the influence of diet on multifactorial characters is compounded by our ignorance of the actual functions of individual genes in polygenic systems. In general, it can be said that an understanding of the expression of any genetic program will require a more thorough elucidation of the basic mechanisms of genetic regulation.

GENETIC REGULATION

DNA, Genes, and Gene Mapping

It is thought that structural genes represent less than 10 percent of the total genome. The remaining genes belong to two categories of repetitive DNA, called highly repetitive and intermediate repetitive. The latter includes DNA coding for stable RNA, and sequences which are transcribed along with the structural genes (10).

Structural loci are often unique, as in the case of the locus which determines the beta chain of hemoglobin. But they may also be lowly repetitive, as in the case of the two non-alletic loci which code for the alpha chain of hemoglobin. These lowly repetitive loci may have minor variations between them, as illustrated by the 136 glycine and 136 alanine cistrons which code for the gamma chain of hemoglobin. Thus it can be said that the genetic information is redundant; and this redundancy, which is responsible for certain variant forms of hemoglobin and also for some isozymes, must have arisen in the course of evolution by successive duplications of "primitive" genes.

It often happens that repetitive loci remain closely linked, as in the case of the HL-A loci which determine histocompatibility in human cell cultures. Other examples would be the clustering of the four loci which code for the heavy chains of immunoglobin G (IgG), and the linkage of the loci which code for the beta, gamma, and delta chains of hemoglobin.

However, there appears to be no linkage and no clustering of genes coding for corresponding enzymes found in different cell fractions (i.e., soluble, mitochondrial, and lysosomal fractions), or of the genes coding for enzymes which participate in a particular metabolic pathway. For instance, the glucose-6-phosphate dehydrogenase (G6PD) locus is on the X chromosome, whereas the

6-phosphogluconate dehydrogenase (6PGD) locus, which codes for an enzyme closely related to G6PD, is on chromosome no. 1. Similarly, the locus for leucocytic pyruvate kinase (PK-3) is on chromosome no. 7, while the locus for hexokinase-1 is on chromosome no. 10. The triosephosphate isomerase (TPI) locus is on chromosome no. 12, whereas the phosphohexose isomerase locus is on chromosome no. 19. Thus there is nothing in the human genome exactly comparable to the operon described in microorganisms (11).

Messenger RNA

Recent advances in molecular biology indicate that the pattern of messenger RNA formation in higher organisms (eucaryotes) differs fundamentally from that observed in bacteria (procaryotes). While bacterial mRNA is synthesized in its final form directly by transcription, it now appears that mRNA in mammalian cells is derived by post-transcriptional modification of larger molecules. These molecules are sometimes called dRNA or HnRNA (heterogeneous nuclear RNA), and are described by Scherrer as _transcription units_ (12). The transcription unit is processed within the nucleus to premRNA, which is composed of one or more messenger sequences, together with true nonsense segments and programming sequences. The programming sequences contain regulatory signals directing, through association with protein, the stabilization of premRNA, its processing and transport to the cytoplasm, and its eventual translation.

This process of intracellular information transfer presents many possibilities for regulation. Several of these have been enunciated by Scherrer in his _cascade regulation_ hypothesis.

Regulation may be defined as a cellular event which allows or affects the accumulation of mRNA molecules. Regulation may occur at the _transcriptional_ and/or the _posttranscriptional_ level - that is, during the maturational process of mRNA and before its attachment to ribosomes. Changes in the output of protein specified by a given amount of mRNA are called _translational modulation_. Finally, there is the possibility that _posttranslational modification_ of the protein, which may occur in several different ways, may play a role in the general process whereby a cell adapts its enzymatic activity to the needs of the moment.

Genetic Regulation in Higher Organisms

As stated above, it appears that in higher organisms there is no system of transcriptional regulation similar to that described by Jacob and Monod in bacteria. This difference may be accounted for by differences in homeostatic mechanisms, in the pattern of mRNA formation, and in the half-life of mRNA. However, there are some features of regulation common to higher organisms and bacteria. In both cases, regulation involves specific proteins which have the capacity to recognize effectors or inhibitors in the intracellular milieu, and also to recognize programming sequences wherever they are located.

This hypothesis is supported by the data obtained from somatic cell hybridization experiments. Somatic cell hybridization involves the fusion of cells derived from two different parental lines, which are unlike one another with respect to some differentiated function. Generally the cells expressing the differentiated function will be derived from a permanent heteroploid cell line, while the cells not expressing it may be fibroblasts from the same or a different species (13). From such fusions, three kinds of results may be obtained:

-- The differentiated function expressed in one parental line may be suppressed in the hybrid. For instance, the synthesis of melanin by Syrian hamster melanoma cells is suppressed in the hybrids formed between these cells and mouse fibroblasts.

-- The differentiated function may be expressed, solely as the result of continuing activity of the relevant genes in the differentiated parent. This is the manner in which the capacity for albumin synthesis is retained in hybrids formed between rat hepatoma cells and mouse fibroblasts.

-- The differentiated function may be expressed, with the relevant genes of both parents contributing to its expression.

These experiments do not provide any unequivocal demonstration of the mechanism of gene regulation in mammalian cells. However, as previously noted, the results do suggest that mammalian cells use diffusible regulator substances to control the expression of many of their differentiated functions.

Levels of Genetic Regulation

The general pattern of suppression of synthesis of lactase at the time of weaning, which is not governed by lactose ingestion, is suggestive of transcriptional regulation, as are other instances in which protein synthesis is switched off or on in the course of development (4). Certain phenomena of cellular differentiation, such as the specialization of lymphocytes to synthesize specific antibodies, and the capacity of certain heterozygotic cells to express the gene of only one member of a pair of chromosomes (as in the immunoglobin Gm haplotype) are also suggestive of regulation at the transcriptional level. However, it must be kept in mind that in heterozygotes the activity of an enzyme is often halved, which would seem to exclude a regulation of synthesis in many cases.

In other instances, variations in the activity of enzymes are probably due to posttranslational modifications. For example, it has been observed sucrase and isomaltase activities are reduced by half when glucose is substituted for sucrose in the diet; reintroduction of sucrose or of fructose allows a quick restoration of the enzymatic activities - a phenomenon which could be due to stabilization of the enzymes by their respective substrates. The increase in activity of the glycolytic enzymes which is induced by glucose is another phenomenon which probably results from posttranslational modulation (14).

Interpretation of results is often difficult, because in certain cases several mechanisms may interplay, as illustrated by the data presented at this conference by Galand, which are suggestive of a posttranscriptional or translational regulation of disaccharidase synthesis induced by cortisone. Other studies have shown that induction of tyrosine aminotransferase by cortisone is controlled at the transcriptional and at the translational levels (6).

Regulator Genes

The use of the term _regulator genes_ has been carefully avoided until now. At times some authors have attempted to interpret many errors of metabolism in terms of regulatory mutations - a concept which has not held up under analysis. Although regulator genes in the narrow sense, as described by Jacob and Monod, do not exist in higher organisms, there must be operators, promoters, and genes coding for programming proteins (16). It has been shown in many cases that loss of enzymatic activity is due to an abnormality of the primary structure of the enzyme. However, in the absence of evidence on this point, it is difficult to determine whether a lack of activity is due to a loss of the capacity to synthesize the enzyme, or to a structural anomaly of the protein itself (as demonstrated by the presence of cross-reacting material).

A loss of synthesis could result from several mechanisms: from partial or total deletion of a locus, as has been suggested in the case of the alpha thalassemias (8, 21); from a nonsense mutation of a structural gene; from an operator-minus (o-) mutation; or from a mutation affecting the translation complex, as suggested by data on induction of beta hemoglobin chains _in vitro_ by beta thalassemic cells (17).

When a missing enzymatic activity can be induced in response to a stimulus, deletion and nonsense mutation can be discarded as possible explanations for the initial loss of activity; however, either of the remaining two explanations might be applicable. Such is the case for the _in vitro_ induction by azauridine of the two enzymes missing in oroticaciduria - orotidine-5'-pyrophosphorylase and orotidine-5'-phosphate decarboxylase (16).

The preceding example focuses attention on the problem of multiple enzymatic defects. Here again, regulatory mutation is the hypothesis of last choice. Consider, for example, the case of sucrose-isomaltose intolerance. Ordinarily, sucrase and isomaltase are bound by different, although highly similar proteins with a high affinity for each other, and brought together as a complex in the brush border of the intestine. Thus creation of an active enzyme complex requires the participation of two discrete loci in addition to the genes which code for the enzymes. In sucrose-isomaltose intolerance, both enzyme activities are missing, and practically no enzyme proteins can be detected in the brush border membrane. These findings could be explained by an o- mutation blocking the transcription of polycistronic mRNA; but a more likely explanation would be that there is an anomaly in the structure of one of the binding proteins, not necessarily the same protein in different cases, which prevents their association, or their attachment to the membrane, or causes the formation of a very unstable complex (18).

Food and Genetic Regulation

It has been observed that amino acid starvation brings about disaggregation of polysomes. It is possible that this phenomenon could have nonspecific consequences for development. It is also possible that the presence or absence of a specific nutrient at a specific stage, particularly in the postnatal period, might have untoward consequences for the regulatory mechanisms controlling the biosynthesis and degradation of this nutrient.

Such a possibility is in keeping with Raihä's remarks at this conference, and has also been suggested in relation to cholesterol metabolism (19). It has been shown that serum cholesterol concentration, which is low at birth, increases soon after birth. Feeding of formulas with fat in the form of butterfat results in serum concentrations nearly equal to those of breast-fed infants, whereas feeding of formulas with fat in the form of vegetable oils results in much lower concentrations.

The lower concentration of cholesterol may have immediate consequences, for example in corticosteroid biosynthesis, if the newborn's ability to synthesize cholesterol is limited. It may also have untoward consequences in the long term, as suggested by experiments in rats: Hahn and

Kirby weaned rats at 15 and 30 days, respectively, and then fed them a cholesterol-free stock diet. At 215 days of age the serum cholesterol concentration of the former group was higher than that of the latter group.

CONCLUDING REMARKS

It is clear that the relationship between food and genetic development is intricate and as yet poorly conceptualized. It is the conviction of this author that further advances in the understanding of the human genome - its organization, its expression, its regulation and the factors involved in regulation, particularly regulatory proteins - will bring new insights in this field.

Acknowledgment

The author would like to thank Professor J. Rey for helpful discussion.

REFERENCES

1. Barthelmess A: Mutagenic substances in the human environment. In Vogel F, Röhrborn G (eds): Chemical Mutagenesis in Mammals and Man. Berlin, Springer Verlag, 1970, p. 69.

2. Adler ID: The problem of caffeine. In Vogel F, Röhrborn G (eds): Chemical Mutagenesis in Mammals and Man. Berlin, Springer Verlag, 1970, p. 383.

3. Tedesco TA, Morrow G III, Mellman WJ: Normal pregnancy and childbirth in a galactosemic woman. J Pediat 81:1159, 1972.

4. Kretchmer N: Lactose and lactase - a historical perspective. Gastroenterology 61:805, 1971.

5. Rey J, Rey F, Schmitz J: Structure of milk glycerides and its species adaptation. Biomed 20:90, 1974.

6. Komrower GM, Lee DH: Long-term follow-up of galactosaemia. Arch Dis Childh 45:367, 1970.

7. Fishler K, Donnell GN, Bergren WR, Koch R: Intellectual and personality development in children with galactosemia. Pediatrics 50:412-419, 1972.

8. Taylor JM et al: Genetic lesion in homozygous α thalassaemia (hydrops fetalis). Nature 251:392, 1974.

9. Harris H, Hopkinson DA, Robson EB: The incidence of rare alleles determining electrophoretic variants: data on 43 enzyme loci in man. Ann Hum Genet Lond 37:237, 1974.

10. Evans HJ: Molecular architecture of human chromosomes. Brit Med Bull 29:196, 1973.

11. Bergsma D (ed): Human gene mapping: birth defects. Original Articles Series XI, No. 2, 1975.

12. Scherrer K: The synthesis of mRNA in animal cells. In: Synthèse Normale et Pathologique des Protéines Chez les Animaux Supérieurs. Colloque INSERM, 1973, p. 113.

13. Davidson RL: Control of expression of differentiated functions in somatic cell hybrids. In: Somatic Cell Hybridization. New York, Raven Press, 1974, p. 131.

14. Schmitz J _et al_: La régulation des enzymes intestinales. Pädiat Fortbildk 42:13. Basel, Praxis, 1975.

15. Darnell FE, Jelinek WR, Molloy GR: Biogenesis of mRNA: genetic regulation in mammalian cells. Science 181:1215-1221, 1973.

16. Dreyfus JC: Contrôle génétique de la biopysynthèse des protéines en pathologie. In: Synthèse Normale et Patholigique des Protéines Chez les Animaux Supérieurs. Paris, Colloque INSERM, 1973, p. 7.

17. Conconi F, Senno L: The molecular defect of the Ferrara β thalassemia. In: Synthèse Normale et Pathologique des Protéines Chez les Animaux Supérieurs. Paris, Colloque INSERM, 1973, p. 329.

18. Schmitz J _et al_: Absence of brush order sucrease-isomaltase complex in congenital sucrose intolerance. Biomed 21:440, 1974.

19. Fomon, SJ: Infant Nutrition. Philadelphia, W.B. Saunders Company, 1974, pp. 172-175.

20. Alpers DH, Kinzie JL: Regulation of small intestinal protein metabolism. Gastroenterology 64:471-496, 1973.

21. Ottolenghi S _et al_: Gene deletion as the cause of α thalassaemia. Nature 251:389, 1974.

22. Vogel F, Röhrborn G: Concluding remarks. In: Vogel F, Röhrborn G (eds): Chemical Mutagenesis in Mammals and Man. Berlin, Springer Verlag, 1970, p. 453.

FOOD AND PSYCHOLOGICAL DEVELOPMENT

Hanuš Papoušek, M. D. Sc.D.

Research Associate, Professor of
Development Psychobiology
Max-Planck Institute for Psychiatry
8-Munich-23, Kraepelinstr. 10
West Germany

In this discussion I would like to invite the audience to look into the world of human infancy to which I have devoted the last two decades in research. Infancy: the stage in life where a steep gain in body weight is observed, where a full and balanced diet is needed to meet the demands of growth, and where the complex preverbal cognitive processes have been shown to be affected by the quality and quantity of nutrition during this period.

For the infant, food's primary purpose is to meet the demands of his body to assure growth; but food and eating also comprise part of the infant's relationship and adaptation, via his central nervous system, to both his internal environment (IER) and his external environment (EER). Adaptation to the external environment utilizes information coming from the telereceptors and depends on a fine and fast analysis of time and space, whereas regulations of the internal environment utilize information from the internal receptors and are less concerned with time and space.

The mutual relationship of these two types of regulation is derived from five aspects:

-- Both types of regulation depend on external environmental factors. Therefore, a change in these factors can lead to disturbances in both regulations simultaneously, e.g., excessive heat, microbial invasion, or intoxication.

-- The state of nutrition (an IER outcome) influences the level of EER in cognitive development and in learning. This aspect is frequently studied in the relation between nutrition and psychological development.

-- Internal metabolic demands activate the behavior oriented toward the external environment. For example, the infant's play changes with the degree of hunger, and this exploratory behavior is then replaced with oral activities.

Koch, a member of our Prague team, analyzed the biorhythmical changes in the level of the infant's learning and found regular fluctuations related to the times of feeding and sleep.

In relation to hunger the infant's performance in conditioning experiments first improved, reached an optimum approximately 60 minutes after feeding, and then decreased (1).

-- Vice versa, the demands resulting from the adaptation in the external environment may activate autonomic processes through IER.

Observation of a four-month-old baby exposed to an attractive but not easily attainable toy demonstrated that simultaneously with the activation of his attention and exploration a larger metabolic reserve was brought to his muscles. For 20 to 30 minutes the baby engaged in very vigorous gymnastics which involved all parts of his body until he became tired.

-- The outcome of adaptive processes involved in EER determines IER. In adults, a decision to reduce food intake and, thus, the body weight, is a typical example. Neurotic disorders resulting from disturbed social interactions are more typical for preverbal infants.

The last two aspects have attracted less attention in the literature than the preceding ones, although they are closely related to the general problems of behavioral regulation. Their potential relation to the origins of psychomatic disorders makes them even more challenging. In my presentation, I want to discuss these last two aspects in particular.

Though there is not a wealth of research in these areas during early development, there is clinical evidence to show that in neonates, disturbance in the exteroceptive behavioral adaptation can lead to disorders in nutrition.

Let me cite two examples.

Gunther described severe refusal of the breast in neonates whose inexperienced mothers caused complete nasal obstruction during the first breast feeding (2).

Jungmannová showed that behavioral maladjustment in neonatal breast feeding leads to the occurrence of rhagades in the maternal breast nipples and to unsuccessful breast feeding (3). The crucial symptom, i.e., excessive sucking on the nipple and chewing on its vulnerable margin.

From our own clinical experience, I could describe dramatic cases of disturbances in the mother-infant interaction resulting in vomiting of any food offered to infants by their own mothers. Experienced nurses had to replace those mothers and carefully re-establish a better mother-infant interaction after full recovery of infants.

A theoretical distinction between behavioral adaptation in the external environment and behavioral regulation of the internal environment was proposed and verified experimentally by Garcia and his co-workers (4).

An illustration, from their work, showing the dichotomy between both systems deals with the rat in a radiation field (5). The rat is able to detect x-rays because they stimulate the olfactory receptors by producing chemical changes in the olfactory mucosa. Thus, a blip of 1 r alone arouses a sleeping rat; a few milligoetgens are enough to signal the rat to avoid subsequent electric shocks; an exposure to 100 r will cause sickness; and an exposure to 1000 r will kill the rat.

If the rat is given a free spatial choice between two clearly marked areas, one exposed to radiation and the other shielded from radiation, it is unable to avoid accumulation of a fatal dose for it keeps entering the exposed area in spite of being able to detect x-rays.

On the other hand, after a single dose of 50 r, the rat will demonstrate distinct alteration of its diet in order to cope with radiation effects. Thus, it will avoid foods consumed prior to, and during exposure.

Obviously, like many other species, a rat can cope with food that produces illness, but not with places that produce illness, even when it can detect the agent producing the illness. Similarly, rats display a persisting aversion for tastes, e.g., for saccharin, which were associated with exposures to 30 r radiation (6).

In our studies of the early development of human behavior, we approached this category of questions when we attempted to conceptualize the regulation of behavioral states and the development of adaptive processes in human infants. In addition to clinical experience showing that neurotic disorders regularly affected food intake and digestion, we also saw that oral activities and autonomic responses were activated with amazing frequency by arousal of attention, exploration, and processes involved in learning or problem solving.

Although experimentation with human infants is, for obvious reasons, very narrowly limited, we were able to draw some conclusions from studies of conditioning with appetitional reinforcement. However, these studies should be introduced with several theoretical comments.

First is the question of the role of cognitive processes in adaptive behavior. Typically, adaptive behavior has been viewed from the point of the most relevant biological determinants related to the survival of individual or of the whole species. From here behavior has been categorized as food seeking or avoiding, social, or reproductive.

Though there are certain common elements in all these individual specific categories and though there is no doubt of the existence of their observable components in the behavioral repertoire, such as orienting responses, exploratory behavior, or manipulatory trials, nevertheless, due to discrepancies in their theoretical interpretations and in the explanation of their survival values, their categorization still remains controversial.

Roughly we may say that these mechanisms in one or another way effect the input of information and its processing on one hand, and the organization of adequate specific responses on the other hand. They differentiate novel stimuli from familiar ones and detect relations among various stimuli or regularities in the complex environmental changes. They discriminate environmental changes which are independent of the organism's own activities from those which result from its activities. They integrate accumulated experience into concepts of individual objects or events.

On the other hand, these mechanisms select adequate behavioral patterns, combine them into complex forms of inborn responses, and enable acquisition of new behavioral responses through learning or problem solving.

Therefore, we view these mechanisms as fundamental cognitive processes which underlie all other specific categories of behavior: appetitional, defensive, social, or reproductive.

We studied the development of some fundamental cognitive processes in infants and found certain relations between their activation and the structure of environmental changes. For example, stimulation with blinking colored lights alone elicited only orienting responses which soon decreased and were habituated, but as soon as we made the same visual stimulation contingent on the infant's own behavior, it elicited much more intensive orienting and exploratory behavior that did not stop before the infant discovered how to manipulate the visual stimulation. Such exploration usually represented very intensive motor activities necessitating an adequate increase in the supply of energy.

On the contrary, exposure to a flood of stimulation or to situations representing too difficult problem situations for the infant caused him to turn away and stop responding as if to protect himself against some nociceptive effects.

Thus, it can be seen that some structural patterns in the environment act as specific stimuli, triggering specific changes in the fundamental cognitive processes, analogous to avoiding, food seeking, social, or reproductive responses in animals.

Figure 1 illustrates our theoretical concept of activation, versus inactivation of the fundamental cognitive processes. It also shows the participation of adaptive autonomic responses. It is, of course, hardly possible to demonstrate clearly what really takes place in the infant's brain when he is exposed to novel stimulation, when he discovers new consequences of his own activity, or when he must cope with a difficult problem situation; but, with infants in particular, we can analyze the indirect evidence of the regulatory processes in the overt behavior. We can clearly observe the activation of locomotor movements, exploratory movements, or oral activities as well as signs of either pleasant or unpleasant feelings and the emission of vocal or facial signals for the social environment (7).

Two points should be emphasized in this connection.

First is the fact that the striking increase of motor activities in difficult learning or problem solving situations is very often accompanied by increased oral activities, such as sucking in the youngest infants, and increased salivation or sucking on fingers or fists from the third or fourth months on, as if the necessity to increase the energetic supply in muscles were accompanied by a need to increase the intake of food. This concept seems to explain the frequent oral activities of infants better than the libidinal concepts, and it certainly allows for better experimental verification.

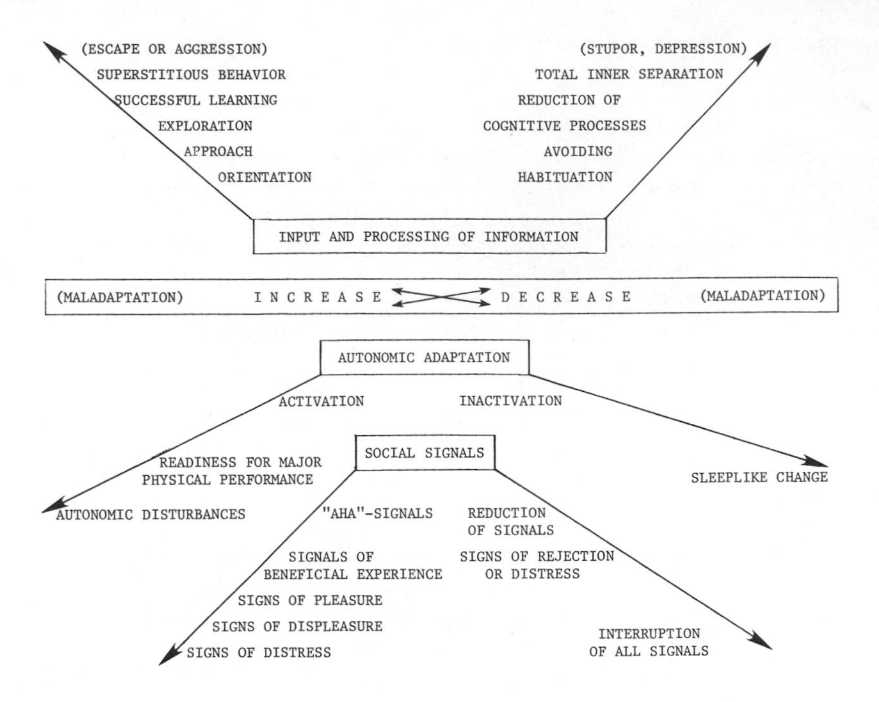

Figure 1 Fundamental cognitive response system

Second, in the accompanying emotional behavior, we found a regular relationship with the processes of learning or problem solving. Phases preceding a satisfactory solution were accompanied with signs of discomfort or distress, whereas the acquisition of correct adaptive responses was followed by signs of pleasure and joy.

This second point supports our ideas about the origin of intrinsic motivation. Next to the pleasure of the rewarding reinforcement in the form of food, there is an additional intrinsic reward in successful learning or problem solving.

From these points of view, everyday feeding reflects the outcomes of both regulatory systems: the satiation of the metabolic demands of the internal environment on one hand, and the course of adaptation to the external environmental circumstances associated with the intake of food in some problematic way.

The outcomes may be concordant at times, and at other times, discrepant. In the former case the pleasure resulting from the satiation of hunger may be multiplied by the pleasure from successful adaptation, e.g., the mastery of skills necessary for drinking from a cup. In the latter case the pleasure derived from food may be overshadowed by difficulties resulting from unfavorable social interactions, e.g., with inept babysitters or other caretakers, in unfavorable family situations, or during hospitalization.

Our studies of early learning abilities using conditioning methods with appetitional reinforcement demonstrated that the motivational aspects of feeding can be subjected to experimental tests.

Healthy infants from birth to six months were taught first to respond to a sound signal with a head turn to the left, and, thus, to receive a small portion of milk. Then they were taught to discriminate two different acoustic signals and turn to the correct side to be rewarded with milk. Such an experimental session replaced one of the regular morning feedings, and during 10 trials making up one such session, the infant obtained his usual amount of milk divided into 10 portions. Thus, in the course of one session, his hunger was gradually fully satiated.

However, when analyzing the course of his conditioned performance in relation to the degree of satiation, we found no significant dependence either in the course of simple conditioning or in the course of conditioned differentiation and its reversals.

We examined this finding even more precisely in four-month-old infants in prolonged sessions with unlimited delivery of milk. We saw that even after complete satiation of hunger, succeeding conditioning signals elicited quick and strong correct head turns and were accompanied by smiling and pleasant vocalization. It was clear that performance of the correct response played a larger role in motivating the infants than the reinforcement with milk and satiation of hunger (8).

In another experiment with a group of 15 infants aged 88 to 201 days, a similar left-right discrimination in head turning was reinforced with different tasting fluids. Sweet milk was used as a reward for left turns, and a mild bitter solution, for right turns. Soon the corresponding two acoustic signals alone elicited concomitant emotional responses appropriate to the kind of reinforcement. When, in this situation, the discrimination was reversed, for some time the infants sucked the bitter solution with the sweet acoustic signal without any sign of displeasure, and refused the sweet milk now presented with the bitter acoustic signal (8).

Our studies led to these conclusions:

-- The motivation of behavior connected with food intake results from both the interoceptive regulatory processes and the cognitive processes involved in the adaptation in external environment.

-- The course of adaptation in the external environment can influence the interoceptive regulation both favorably and unfavorably.

-- The interrelation between interoceptive and exteroceptive regulations can be detected in early infancy.

With the knowledge that experiences in infancy can significantly influence subsequent development, we should also pay attention to the practical implications of these conclusions.

Parental experiences can demonstrate two unfavorable conditions: one in which severe disorders in the external environmental adaptation constantly disturb the interoceptive regulation, thus eliciting psychosomatic diseases or their analogies; and the opposite, in which feeding occurs in boring, monotonous, and never problematic conditions, thereby preventing additional pleasure which could be derived from successful exteroceptive adaptation.

The latter consequence is less well known, but it is perhaps just as dangerous today when even the infant's world is polluted with ready-made, instant, and monotonous products and second-hand images of life and nature.

In addition to hunger and malnutrition, we have here presented our theoretical interpretation of two additional problems relating to food and feeding which may also adversely affect early psychological development.

REFERENCES

1. Koch J: The change of conditioned orienting reactions in five month old infants through phase shift of partial biorhythms. Hum Dev 11:124-137, 1968.

2. Gunther M in Foss BM (ed): Determinants of Infant Behaviour. London, Methuen, 1961.

3. Papoušek H, Jungmannová C, in Kubát K (ed): Péče o novorozence. Praha, Státni zdravotnické nakladatelství, 1967.

4. Garcia J, Ervin FR in Wortis J (ed): Advances in Biological Psychiatry. New York, Plenum, 1968.

5. Garcia J, Koelling RA in Beidler LM (ed): Handbook of Sensory Physiology, V. 4, The Chemical Sense. Berlin, Springer Verlag, 1971.

6. Rzoska J. Brit J Anim Behav 1:128, 1953.

7. Papoušek H, Papoušek M in Nissen R, Strunk P (eds): Seelische Fehlentwicklung im Kindesalter und Gesellschaftsstruktur. Neuwied, Luchterhand, 1974.

8. Papoušek, H in Stevenson HW, Hess EH, Rheingold LH (eds): Early Behavior: Comparative and Developmental Approaches. New York, Wiley, 1967.

somatic growth spurt period is not known to be associated with any particular vulnerability to long-term growth restriction.

For those of us interested in humans, the relevance of these animal studies must now be discussed. It is obvious that our knowledge of the effects of undernutrition on the structure of the developing brain - the deficit and distortion pathology - could only have been observed in studies performed with experimental animals, where the investigator has precise control over the experimental conditions and the facility for killing and examining the brain at any time. Transferring the findings to our own species by inference, however, requires certain strict rules of cross-species extrapolation, and these have very often been disobeyed. The main problem has been that although the sequence of brain developmental events is the same from one animal species to another, and although the unit components of the brains are virtually identical, the brain growth spurt is differently timed in relation to birth. Thus a rat is born before the period begins. The guinea-pig completes its brain growth spurt as a fetus before it is born. Therefore, since developing brain vulnerability is related to the <u>events</u> of brain growth, and not to age from birth, it will be necessary to know the timing of these events in man before we can identify the human period of vulnerability.

We have recently reported a study of human brain development (4) which shows, contrary to earlier findings, that the brain growth spurt in the human is mainly a postnatal affair. It is true that it begins a mid-gestation, but it lasts at least until the second birthday, and probably beyond. Now, since fetal growth is not known to be affected by maternal malnutrition until the beginning of the last trimester of pregnancy, it can be surmised that only the last three months of fetal life of the brain can be affected out of a total growth-spurt period of well over two years; and in these circumstances, as we know from animal models, the fetal human brain will not be irretrievably affected. The animal experiments tell us that undernutrition need not be severe but it has to be prolonged, lasting throughout the major part of the brain growth spurt, if it is to have detectable effects on the physical brain. And although this need only be three

FOOD AND HUMAN BRAIN DEVELOPMENT

John Dobbing, B.Sc., M. D.

Professor of Child Growth and Development
Department of Child Health
University of Manchester
Manchester, England

The question of whether nutrition affects the full achievement of human intellect is quite old, but it is only in the last decade that it has begun to be seriously studied. In the early part of this century, when starvation was nearer to the North American and European scientist in his own daily life, there were many excellent animal studies by well-known nutritionists to determine whether starvation affected the brain; the clear conclusion emerged that it did not. Animals were occasionally found which had starved to death, and experimental rats were kept on starvation diets; although it was easy to see how nearly all the organs and tissues of the body were profoundly affected, the most dramatic exception was the brain. Thus was born the doctrine of "brain sparing," and by about 1925 most research in this area stopped. There seemed little professional future in it.

More recently we ourselves starved twelve adult rats for a period of weeks by keeping them on a diet of sucrose and water, and when the first one died we killed the rest. They were reduced to one half their former weight, and there were all the known metabolic distortions of extreme inanition; but their brains were quite without any sign of change, either in weight or in their detailed chemical composition.

The discovery which reopened the whole question a little more than ten years ago (1) was that the apparent immunity of the <u>adult</u> brain to starvation was not shared by the *developing* brain, which, at certain vulnerable periods of development could be adversely affected, and permanently so, by even mild dietary restriction. The earlier experimenters had also looked at growing rats. However, because they had confined their attention to the weanling animal, and because the rat has almost completed its period of vulnerability by the time of weaning, these investigators never observed the full impact of malnutrition on the developing brain.

At that time my colleagues and I were working on the structural and metabolic features of myelination, which occurs in the younger, suckling rat. The question we asked was whether the myelin sheath (which is remarkably stable in many ways once it has formed) may yet be vulnerable to nutritional adversity during its period of formation; and so it turned out to be. It then gradually became clear that many other features of brain development were also vulnerable at this critical time, and the concept of vulnerable periods in the developing brain began to be elaborated.

At about this same time the classical human field studies of Cravioto and Robles (2) and others were under way, pointing to the conclusion that later intellectual development appeared to be impaired in children malnourished in their first two years of life.

In its earliest form the vulnerable period hypothesis stated that the developing brain was most vulnerable to restricted nutrition at the time it was growing fastest - the *brain growth spurt* - and it was subsequently shown that the vulnerability was so great at that time that even quite mild undernutrition, correctly timed, produced lasting deficits and distortions from which the brain could never recover. This is in marked contrast with the earlier finding of complete sparing of the developed brain in the older animal.

The lasting changes which undernutrition produces during the brain growth spurt will only be briefly enumerated here since they have been described many times before (3). It is important to understand, however, that we still do not know whether any of these changes is really important to intellectual function. We can only assume that

they may be, but since we still know nothing of the physical basis of intellect, it would be irresponsible to conclude that the changes in the brain which we can easily measure are necessarily themselves of ultimate significance.

Undernutrition during the brain growth spurt (but not later) permanently reduces the size of the brain a little more than it reduces body size; but it does not affect the brain uniformly. The cerebellum, for example, is more reduced than the rest of the brain, its differential vulnerability being related to its higher rate of growth. Even the cerebellum is not uniformly reduced, there being a specific deficit in those cerebellar neurons which multiply during the brain growth spurt period. There permanent cerebellar changes are certainly related to later neurological function, since the affected animals remain permanently clumsy. If there were a neuroanatomical description of intellect as there is for clumsiness, this would be a much easier subject. Unfortunately, we can only make assumptions by analogy, and they may be wrong.

So it can already be seen that the pathology of the brain produced at this time by undernutrition is not one of destruction. This is not a "lesion" pathology of the traditional kind, often referred to colloquially as 'brain damage'. It is a growth pathology resulting in permanent deficits and distortions, without any areas of focal damage.

Since the discovery of the vulnerability associated with the brain growth spurt period, other periods of brain vulnerability have been identified. In general, however, these are only of theoretical significance for indigent human society. The period of the brain growth spurt continues to be recognized as a time of highly significant vulnerability, although many details of the mechanism remain to be worked out. Before leaving the experimental evidence derived from animal studies, it m be useful to mention that brain growth spurt period is associated with vulnerability to restriction of somati growth as well as brain growth. This is an unexpecte finding, since the somatic growth spurt is known to b after the brain growth spurt is completed; moreover,

weeks in a fast-growing species like the rat, it must, by analogy be at least two years in the human. Thus the small proportion of the vulnerable period represented by human fetal life is almost certainly too short a time to produce irrecoverable changes in the developing human brain. In real life conditions, however, poor fetal growth is nearly always followed by poor growth in the postnatal years; and in those circumstances the fetal and postnatal effects will be additive. In other words, it is probably correct to say that fetal growth retardation due to maternal undernutrition need not be of permanent consequence; but it will contribute significantly to ultimate disability unless the child is liberated at birth from growth-restricting influences.

One of the problems of investigating this matter directly in humans is that the illiterate and non-numerate mothers who have to be studied are unable to give an accurate history of the length of their pregnancy. The brain is growing fast at the time of birth, and an inaccuracy of one or two weeks in assessing gestational age destroys the usefulness of brain measurements. There is, alas, no shortage of dead babies' brains from malnourished mothers, but they have not hitherto been reliably examined, partly because of this difficulty in dating them.

We have recently identified a grossly undernourished society in which recollection of menstrual history appears to be nearly perfect. The reasons are that the religious, social, and sexual customs of the women in this particular society impose an absolute recall of each menstrual period which is identified by named days on the religious calendar. There are, in fact, several religions whose adherents have sexual taboos numerically associated with the days of menstruation, and who may be (for example) forbidden entry into a place of devotion if they are menstruating. We have been able to obtain about one hundred and twenty complete brains from babies in such a disadvantaged society, ranging in age from about 27 fetal weeks until full term. These have been subjected to quantitative analysis. Preliminary results do indeed show a measurable quantitative deficit in several indices of brain structure by the time of birth. Compared with British fetuses there are no measurable deficits at 27 weeks, as might have been expected, since nutritional fetal growth restriction only occurs in the third trimester. But throughout the last three months there is a steady falling off in weight of the various brain regions, in their cell number, and myelination, which are the traditional anthropometric measurements of the growing brain. By the time of full-term birth these quantities are quite clearly reduced in the babies of undernourished mothers. Nonetheless, the size of the reduction is almost certainly potentially recoverable, especially since five-sixths, or six-sevenths of the brain growth spurt period in humans is postnatal and is yet to come. The remaining period is one in which we have a theoretically ample opportunity to promote good brain growth and rescue a malnourished fetus from its disadvantage at birth. Obviously, however, we have not yet devised a way of ensuring this growth promotion, even in a world whose technical achievements are formidable.

A very distinguished developmental neurobiologist in the field of mental retardation recently asked me to justify the time and money presently devoted to the study of the effects of malnutrition on the developing brain. His point was that we could only end up with the not-very-helpful conclusion that *we should feed children*! My reply was that there were two areas of our subject in which we still require research, even if we restrict our endeavour to that which has direct social purpose. I cannot discuss whether it is "useful" to investigate the detailed mechanisms by which malnutrition restricts and distorts brain growth. My own prejudice is that it is *not* useful. However, it is of immense practical and political importance to know whether there really are well defined periods of brain vulnerability; and, secondly, it would be equally important if it could be shown that there were specific components of the diet which were especially required by the growing brain.

There now seems to be no doubt that the brain has reasonably defined periods of vulnerability, as has already been discussed. Furthermore these periods seem to be unique opportunities for the brain to grow properly. If conditions are not good at these times, there seems to be no possibility for subsequent recovery. In short, the period from about thirty weeks of human gestation until at least the second birthday is one in which we must actively promote good brain growth, at least in the present state of our knowledge. Presumably this could sometimes mean that other age groups in childhood might have to receive a lower priority, at least in times of temporary, acute famine.

The question of whether there are specific nutriments for promoting brain growth is also still highly speculative. My own hunch is that there are not. My grandmother used to believe that fish was a good food for the brain. Indeed, Bertie Wooster used to feed fish to his excellent butler, Jeeves, whenever Jeeves had a problem to solve of any intellectual difficulty. There was also a time when it was first discovered that the brain contained "glycero-phosphates," and this led enterprising manufacturers to sell preparations containing them, for the supposed welfare of the brain. The somewhat naive concept that the organs of the body are each directly connected by a tube with the mouth, so that substances which are eaten are directly incorporated into them, is not yet dead. Two distinguished propagandists for breast feeding support their case in a very recent paper (5) with the proposition that human milk contains a number of specified substances which the developing human brain "requires." Several others in the field of lipid neurochemistry have proclaimed that certain essential fatty acids, which are said to be deficient in our domesticated meats, are important to the growth of intellect, and it has even been said that we are in danger of "breeding a race of morons" if we ignore this need. Even our obsession with lack of protein as a specific, prevalent limiting factor for good growth (an idea which is currently being widely challenged) has impinged on the subject of proper intellectual achievement.

My own prejudices are against all such propositions. I have been quite unable to discover any evidence in the experimental literature that lack of protein has any specific role in producing physical deficits and distortions in the brain, or that it is specifically related to ultimate failure of intellectual development. With regard to the role of fatty acids I am much more impressed by the abundant evidence that the brain is able to regulate its own fatty-acid composition, even in the face of severe alterations of fatty-acid intake during development.

Developmental neurochemists have in general found that the brain possesses quite extraordinary self-regulatory powers in extracting the required substances in correct quantity from the circulating blood and synthesizing the components of its own structure. Large fluctuations in availability of precursor substances from the blood are without effect on this process.

I believe the pathogenesis of brain deficits and distortions during development is of quite a different kind, and may well be due to the mere fact of growth restriction, rather than to the consequences of any specific dietary deficiency. When growing children are growth-retarded by underfeeding they are not simply miniaturized. Their bodily configuration is <u>distorted</u>. For example, their length is less affected than their weight; their heads (not only their brains) are less affected than their bodies; their teeth and tendons than their livers. This is due to the differential susceptibilities of organs and tissues to growth restriction; and this, in turn, is related to the differing times during development when organs grow, as well as to their varying metabolic properties. Thus growth restriction at one age may have results different from that at another. Perhaps the same concept can be applied to the developing brain. Like the body as a whole, the brain is far from homogeneous. Its areas pass through growth spurts at different times and at different rates. It might, therefore, be expected that mere growth restriction at any one time will affect those parts which are growing fastest at that time; and if it is true that recovery is more restricted in the brain, the distortion so produced may be permanent. There is already some evidence for this hypothesis. The best evidence comes from the differential susceptibility of the cerebellum referred to above, which is related to its more rapid rate of growth. This phenomenon in no way depends on the <u>quality</u> of the inadequate diet, merely on its quantity.

The title of this symposium, "<u>Food</u>, Man, Society" is, therefore, highly appropriate to my subject; especially where it refers to <u>Food</u>, and refrains from specifying proteins, fatty acids, glycerophosphates, or fish.

In summary it is essential, in discussing such an important human problem as the effects of malnutrition on human intellect, that we neither exaggerate nor minimize our evidence. Malnutrition does not cause "mental retardation," if by mental retardation we mean such conditions as Down's syndrome, severe microcephaly, and idiocy. At the same time the old doctrine of total resistance of the brain is wrong, and we, therefore, have to worry. My own belief is that nutritional growth restriction does have a specific part to play in lowering human achievement, along with all the other environmental disadvantages which so often accompany it. Perhaps no single disadvantage plays a major part, and perhaps humans have a great capacity to compensate for one disadvantage by advantage in another direction. Recent evidence that permanent intellectual deficit only occurs in malnourished children where the non-nutritional environment of the child is also bad can perhaps be explained by this great facility for compensation. However, if no single disadvantage, such as malnutrition, plays a decisive part in reducing human achievement, we are not entitled to conclude that such a disadvantage is unimportant. The end result depends on the quality of all the separate components, and we must look after them all (6).

REFERENCES

1. Dobbing J and Kersley JB: The influence of early nutrition on brain cholesterol accumulation during growth. J Physiol 166:34 p, 1963.

2. Cravioto J, Robles B: Evolution of adaptive and motor behavior during rehabilitation from kwashiorkor. Amer J Orthopsychiat 35:449, 1954.

3. Dobbing J, Smart J: Vulnerability of developing brain and behaviour. Brit Med Bull 30:164, 1974.

4. Dobbing J, Sands J: Quantitative growth and development of human brain. Arch Dis Child 48:757, 1973.

5. Jelliffe DB, Jelliffe FFP: Human milk, nutrition, and the world resource crisis. Science 188:557, 1975.

6. Dobbing J, In, Davis JA, Dobbing J (eds): Later development of the brain and its vulnerability. Scientific Foundations of Paediatrics. Philadelphia, Neinemann, London, and Saunders, 1974.

NUTRITION AND DEVELOPMENT

E. Rossi, M. D.
N. Herschkowitz, M. D.

Universitat Bern
Freiburgstrasse 23
3008 Berne, Switzerland

Understanding the many facets of the relationship between nutrition and development is probably one of the most compelling challenges in the world today. We understand development to include both physical growth and psychosomatic maturation. The high probability that development is significantly affected by nutrition gives this topic a high priority.

Human growth is the result of the impingement of environmental factors on the genetic constitution. Some of the sociocultural factors influencing the genotype leading to the phenotype are the home and community health standards, the family life style, the socioeconomic environment, infections, nutrition, and availability of medical care. Of all of these factors, nutrition is the most prominent, affecting either directly or indirectly the future development of the child. It is possible to study the correlation between growth and nutrition by investigating the deviations of normal growth. In developed countries overnutrition and/or metabolic disorders are the main nutritional problems in childhood, though undernutrition also exists; but in underdeveloped countries it is the hypocaloric, hypoproteic nutrition which demands our attention. The investigation of these problems needs an eclectic approach to cover their various aspects. Therefore, malnutrition is a problem concerning not only physicians, nutritionists, food technologists, biologists, and biochemists; but it also concerns politicians, sociologists, teachers, and welfare workers. The

World Health Organization (WHO) estimates show that ten percent of the world population and two thirds of all preschool children are hypocaloric and hypoproteic. Using clinical criteria, at the most, only seven percent of all cases of malnutrition can be diagnosed. A clinical diagnosis is possible only when clear clinical signs of kwashiorkor or marasmus are present. Because the rest often escape clinical detection, we urgently need methods to identify them. Of those which cannot be diagnosed clinically, and must, therefore, be diagnosed biochemically and biologically, two thirds show moderate and one third show mild malnutrition. Based on this knowledge, the comparison of our problem with a floating iceberg is correct, for we are only able to see the peak of this whole mountain.

In 1974 Rao (1) published a statistical evaluation of the distribution of malnutrition among preschool children. Based on this data the number of children in the world suffering from malnutrition today is about 400 million.

The criteria used today to evaluate the effects of nutrition on development fall into three main categories: physical and mental development, biological and biochemical development, and psychosocial maturation. The fact that so many undernourished children show a possible impairment in one of these three parameters, and most frequently in mental performance, gives this problem great priority. Based on this premise, the gap between developed and underdeveloped countries will become even wider if a nutritional balance at an acceptable level is not soon achieved.

Malnutrition can be manifested at different levels of cell functions: on messenger RNA formation, protein synthesis, turnover of metabolites, the formation of structures, and the development of physiological functions. Finally, an effect on the behavior will be noted.

A primitive organism, which uses a lot of its genome to synthesize essential nutrients, will need little nutrition from the exogenous sources. However, less of the genome will be used for differentiation processes. Man, in contrast, uses little of his genome to synthesize nutrients;

therefore, the organism depends on the availability of exogenous essential nutrients. The advantage of this situation is that a high level of differentiation can be achieved, but only in the presence of these essential nutrients. In absence or imbalance of the essential food components, the differentiation of the organism can be severely affected.

As Widdowson and McCance (2) state, undernutrition reduces the rate of growth and the total phase of growth is prolonged. However, this prolongation does not correct the decreased growth rate. The result is simply a smaller adult person (3).

It is accepted today that the effect of malnutrition on organ development is greatest when there is rapid growth connected with metabolic activity. Dobbing defines this as a vulnerable phase (4). Thus, the effects of malnutrition on development depend on the time when malnutrition is experienced, its severity, and its duration. These factors determine whether the damage will be reversible or irreversible. Each human being has his own genetic make-up which is basically responsible for his growth and development. Infants of small mothers are born small. However, if they happen to have tall fathers, their growth in height and weight may be accelerated, thus allowing them to achieve the genetic limit of their growth. Infants whose height and weight fall below their genetic potentialities because of malnutrition can show "catch-up growth," an accelerated growth rate resulting from normal nutrition following a period of undernutrition. This effect can be particularly pronounced when the duration of undernutrition has been short.

In contrast, Widdowson and McCance (2) demonstrated in an experiment with rats that long duration of malnutrition, occurring between the third and twelfth postnatal week, could not be followed by a catch-up growth when the rats were later placed under normal nutrition. This research is illustrative of the effect of undernutrition coincident with a vulnerable phase of development. These animals will not reach their genetically determined growth.

The effects of undernutrition on development in man can best be studied using clinical means to measure height and to estimate bone age. The activity of the DNA polymerase complex is decreased by malnutrition, thus making possible an estimation of the time during development when malnutrition occurred. After a period of extreme malnutrition all these aforementioned parameters can be only partially corrected by normalization of nutrition. Rosso and Winick (5), in 1974, conducted a study in Guatemala of food supplements given to undernourished pregnant women. This supplement resulted in an average increase in the weight of the newborn baby of 400 g.

Well documented studies of the growth rate of boys in Stuttgart, Germany, between 1910 and 1950 showed a decrease during the periods of World War I and World War II. Though the cause of the decreased rate of growth was not clear, the assumption was that undernutrition played a significant role.

A study of restricted nutrition during gestation of the rat showed that newborn rats, compared with rats fed normal diets, had decreased body weight, brain weight, brain DNA, and protein.

Hohenauer (6) studied the development of twins who had a difference in birth weight of more than 300 g, though they lived in the same environment. It can be assumed that the reason for the weight difference was a malnutrition of one of the twins during pregnancy. The comparison of several parameters of development was undertaken at a median age of 8 1/2 years. The researcher reported statistically significant differences in head circumference, arm circumference, I.Q., and school achievement disfavoring the smaller of the twins. However, there were no statistically significant differences in body height, weight, skinfold of the triceps, and bone age. On this basis the researcher concluded that the prenatal malnutrition had a lasting effect mainly on the mental performance of the smaller twin.

In summary, the effect of malnutrition on growth depends mainly on three factors:

-- the duration of malnutrition
-- the severity
-- the phase in development when malnutrition occurred.

Malnutrition is not the only factor affecting growth. Infections, both directly and indirectly, can affect the developing organism. The combined effects of malnutrition and infection on growth are well documented today both in animals and in humans. The synchronized action of malnutrition and infection affects growth more severely than malnutrition or infection acting alone; one aggravates the other.

A statistically significant example in this connection is the lethal effect of measles in underdeveloped as compared with developed countries. Mexico has a 180 times higher mortality to this disease than do developed, industrialized countries; Guatemala, 280 times higher; and Ecuador, 488 times higher (7). This effect does not seem to be due to a specific viral agent in these countries because natives living in better economic conditions, or Europeans living in these countries show a mortality to this disease similar to that found in the industrialized nations.

According to Latham (8), infections activate adrenal function and mobilize amino acids from the tissue, especially from muscle. Part of these amino acids will be lost as nitrogen in the urine and, thus, a depletion of body protein will result.

Secondary factors exacerbate this basic mechanism, particularly anorexia, vomiting, hypoprotein diet, and inadequate medical treatment, resulting in a shift in the metabolic balance to a degree that the damage may be irreversible. The consequence of all of these factors taken together can be an arrest of body growth. Such an effect has been noted in children suffering from infectious diseases. The problem worsens, becoming a vicious circle, when such children have repeated infections complicated by substandard living conditions. Gastrointestinal disturbances have a more dramatic clinical course in children with marasmus or kwashiorkor than in normally nourished children. So-called weanling diarrhea is an example of such an infection (9).

In summary, this concept emerges: Infections may affect the nutritional state of the child, leading to malnutrition. Malnutrition will result in the loss of protein. The decreased absorption of proteins will increase the degree of malnutrition, and this state is by itself a disposition for further infections. Scrimshaw et al. demonstrated both in animal experiments and in human cases that dietary deficiency reduces the resistance to infections (10).

Children with kwashiorkor are not capable of forming antibodies against typhoid vaccine or diptheria toxoid. This deficiency can be normalized by supplementation of protein foods. The undernourished child in such conditions is known to show a decreased intracellular antibacterial activity. This is connected with decreased adenosine triphosphate levels, and a reduced activity of adenin-oxidase in leukocytes of malnourished children results in a decreased phagocytic activity (11). The decrease of serum transferrin and the diminished integrity of the skin and the mucoids will facilitate the invasion of infectious agents (12).

In addition to these factors there may further be a lack of vitamins, e.g., vitamin C, leading to increased vascular fragility; riboflavin deficiency resulting in angular stomatitis; atrophia of intestinal villi, which will result in a diminution of intestinal absorption; and especially severe xerophthalmia, if there is also a lack of Vitamin E (13). Increased mortality due to infectious diseases is found particularly in children under five years of age. Statistics organized by the Pan American Health Organization and WHO (14) have demonstrated that in 35,000 children under five years of age dying of infectious diseases, 57 percent showed signs of malnutrition, and this malnutrition was directly or indirectly the cause of death.

Supplementation of nutrients to children in Columbia in a wide experimental design showed that under these circumstances morbidity and mortality were significantly decreased.

Of great interest is a study carried out by Taylor and DeSweemer (15) in Rangwal, Punjab, India. The population was divided into four groups. Group one received supplemental nutrition and medical care. The results with regard to morbidity and mortality based on infectious disease were optimal in this group. The second group received only alimental supplementation. In this group there was still a good result. The third group received only medical care. In this group the result was insufficient.

The situation becomes more severe when deceased children are replaced with adopted children. This worsens the whole socioeconomic situation. Unfortunately, children who survive the double effect of infection and malnutrition may exhibit retarded psychomotor development. During the years following this very difficult situation within the family there may also be a lack of positive stimulation for the growing child.

In 1974 Canosa (16) published the results of an investigation by the Institute of Nutrition of Central America and Panama (INCAP) in Guatemala. With respect to physical growth matching among preschool children, four anthropometric measures were applied: weight, height, arm circumference, and tricipital skinfold thickness. Other parameters were the mental development and the socio-cultural status.

The author presented comparative values for height and weight between Guatemalan and United States preschool children. The Guatemalan sample included 2,800 children, representing 92 percent of this population group from eight rural communities. The standards of comparison for both measurements were the Iowa growth curves.

The growth rate in height during the first three months of life is the same for both an urban healthy sample and a rural sample. After three months, however, there is a marked deceleration of growth rate among rural children, lasting until about 30 months of age. Beyond 30 months they begin to recover and by 60 months their rate for height is practically the same as that of urban children.

There were significant differences between the cortical thickness as measured in the second metacarpal, in comparison with norms of the Fels Research Institute for a North American population.

A pilot study was conducted among 20 children who had been clinically diagnosed as malnourished at some point in early childhood, but who had been nutritionally rehabilitated at the time of testing. Height was also taken into consideration for the classification of malnutrition. The heights of the children in the sample were 14 percent or more below normal with regard to age. The results obtained were compared with those of ten of their siblings, who had no clinical history of malnutrition and whose height-for-age ratio showed a deficit between zero and 10 percent. Considering these results Canosa (16) concluded:

> In the populations under study there are marked nutritional deficiencies starting with pregnant women and early childhood. These deficiencies are reflected in placental alterations, poor and distorted physical growth, and high mortality rates.
>
> Children recovered from severe PCM performed less well in psychological tests than their siblings, pointing to the possibility that severe malnutrition within the same social environment can produce changes in mental development.
>
> Long-term prospective studies are necessary to obtain data with which to demonstrate and quantitate the relative importance of the effects of sociocultural and nutritional factors on mental development in developing countries.

In the fight for appropriate nutrition it is reasonable to try to set priorities. The national development will be different in industrialized versus unindustrialized countries. It is reasonable to ask for the underdeveloped countries if it is more important today to have a forced industrialization, an increased economical productivity, a high per capita income, and large, overspecialized medical centers; or if it is preferable to work for improvements of hygienic conditions, the achievement of equilibrated nutrition, the support of agricultural development, the prevention of infectious diseases, the development of dispensaries, and rural medical care. In a recent meeting in India it was stated that today the health situation of children living in the rural areas of that country is worse than in the ghettos of the cities. This is one of the consequences of the dispersion of the population which makes it impossible to achieve critical distribution of doctors in order to deliver health care to everyone. It is easier to get medical care in a ghetto or slum than it is in rural areas.

In order to diminish the incidence of malnutrition in the world, several actions should be taken. Simple ones are encouragement of breast feeding, intensive education for family planning, and intensification of an overall prophylaxis of infectious diseases. In nutrition vitamins must be supplemented; salt must be iodized; serum iron values should be normalized in order to prevent hyposideremia from having an anorexic effect which would lead to retardation of growth. Compensation for lack of proteins can usually be accomplished with supplementation in the form of cereals whose protein component is high. However, of all the frequently used prophylactic measures, breast feeding is the one most readily available and can achieve satisfactory nutrition for infants in the first months of life.

Though mother's milk is generally accepted as the best nourishment for infants, it has additional advantages: the mother's immune globulin protects the child from infections during early infancy; anovulatory effect of prolactin protects the mother from a new pregnancy too soon; there is no more economical form of nourishment: and finally, the mutually satisfying psychological aspect of a mother feeding her baby enhances harmonious development of the child-mother relationship.

Coincidental with the improvement of socioeconomic conditions the percentage of mothers breast feeding their babies has declined in certain countries, though in Northern European countries, probably due to the effect of promotion of the advantages of breast feeding among middle class families, the percentage of breast feeding there has increased.

In conclusion, malnutrition combined with infections may be the most important factors leading to abnormal psychosomatic development. The solution of this problem is to be found in the multidisciplinary cooperative approach which is espoused by this Organization for the Study of Human Development. People whose main concern is optimal human development are the ones to whom this challenge has been given.

REFERENCES

1. Rao AS: Community prevalence of PCM in various countries. WHO Chron 28:172, 1974.

2. Widdowson EM, McCance RA: A review: new thoughts on growth. Pediat Res 9:154, 1975.

3. Malcolm LA: Growth retardation in a New Guinea boarding school and its response to supplementary feeding. Brit J Nutr 24:297, 1970.

4. Dobbing J: Vulnerable periods of brain development. In: Lipids, Malnutrition and the Developing Brain, A Ciba Foundation Symposium. Amsterdam-London-New York, Associated Scientific Publishers, 1972, p. 9.

5. Rosso P, Winick M: Intrauterine growth retardation. A new systematic approach based on the clinical and biochemical characteristics of this condition. J Perinat Med 2:147, 1974.

6. Honenauer L: Studien zur intrauterinen Dystrophie. Pädiat Pädol 6:17, 1971.

7. Morley D: Paediatric Priorities in the Developing World. London, Butterworth, 1973.

8. Latham MC: Diet and infection in relation to malnutrition in the United States. NY State J Med 70:558, 1970.

9. Gordon JE, Chitkara ID, Wyon JB: Weanling diarrhea. Amer J Med Sci 245:345, 1963.

10. Scrimshaw NS, Taylor CE, Gordon JE: Interaction of nutrition and infection. Wld Hlth Monogr Ser 57:1968.

11. Selvaraj RJ, Bhat KS: Phagocytosis and leucocyte enzymes in protein-calorie malnutrition. Biochem J 127:255, 1972.

12. Coovadia HM et al.: An evaluation of factors associated with the depression of immunity in malnutrition and in measles. Amer J Clin Nutr 27: 665, 1974.

13. Latham MC: Nutrition and infection in national development. Science 188:561, 1975.

14. Puffer RR, Serrano CV: Patterns of mortality in childhood. Pan American Health Organization - WHO, Sci Publ 262, 1973.

15. Taylor CE, DeSweemer C: Nutrition and infection in: Rechcigl M jr (Ed): Food, Nutrition and Health. World Rev of Nutr and Dietics 16:203. Basel, Karger, 1973.

16. Canosa CA: Malnutrition and mental development in underdeveloped countries. Padiat Fortbild Prax 38:45, 1974.

17. Greulich WW, Pyle SI: Radiographic Atlas of Skeletal Development of the Hand and Wrist. Stanford, California, Stanford University Press and London, Oxford University Press, 2nd edn, 1966.

18. Jelliffe DB, Jelliffe EFP: Human milk, nutrition, and the world resource crisis. Science 188:557, 1975.

BIOCHEMICAL DEVELOPMENT AND NUTRITION OF THE NEWBORN

Niels C. R. Räihä, M. D.

Department of Obstetrics and Gynecology
Helsinki University Central Hospital
Helsinki, Finland

INTRODUCTION

Considerable advances in obstetric and neonatal care during recent years have markedly increased the survival of low birth weight infants. We are, however, still confronted with the important question of the quality of the long-term outcome of the babies who survive. In the past many follow-up studies on small, prematurely born infants have revealed a high frequency of both physical and mental handicaps, and the general view of the long-term outcome has been fairly pessimistic. It is obvious that most of our efforts should be concentrated on the prevention of premature labor, but before the prematures can be completely eliminated we are faced with the problems of intensive care of these infants, and there is a very real possibility that outcome is to a great extent related to postnatal care and nursery routines. There is increasing evidence that feeding regimens used may be of importance for the development of the low birth weight infant.

Brain growth normally proceeds at a very rapid rate during the last weeks of intrauterine life in the human (1) and it has been proposed that this growth is most vulnerable to disturbances in the supply of nutrients (2). In this regard an overzealous effort of feeding and administration of nutrients may be as harmful as underfeeding.

BIOCHEMICAL IMMATURITY

For some time our laboratory has been interested in studying the development and regulation of enzymes of amino acid and nitrogen metabolism in the human fetus and neonate. These studies and studies in other laboratories have made it clear that the prematurely born, low birth weight infant has many metabolic immaturities which are primarily caused by enzymatic deficiencies. This is particularly evident in amino acid metabolism and, in a sense, transient inborn errors of metabolism are present.

These metabolic immaturities effect both amino acid synthesis and degradation. In the adult mammal, including man, most of the cystine is derived from methionine via the transulfuration pathway. We have recently shown that the last enzyme in this pathway, cystathionase, has no, or very low activity in the human fetal liver and does not appear in the premature infant until some time after birth (3). Thus, cystine should be considered to be an essential amino acid for the premature infant. In the light of current interest in the effects of nutritional deprivation on brain development the nutritional requirement of cystine and possibly other "essential" amino acids in the premature infant is of primary importance. The work of Winick and others showing the debilitating effects of gross protein deprivation has mainly been performed on rats and the human data are related only to very severe malnutrition (2). Protein synthesis in all cells is dependent upon the presence in the right proportion of all amino acids; the complete absence of any amino acid will prevent protein synthesis. When the supply of amino acids is deficient the amino acid which has the lowest concentration becomes rate-limiting for protein synthesis and growth may become impaired. It is, thus, essential that the premature infant not only receive sufficient amounts of protein but also the right quality of protein, containing sufficient amounts of the amino acids which are essential.

On the other hand, the enzymatic capacity for amino acid degradation is also impaired in the low birth weight infant. An example of this is the metabolism of the phenolic amino acids, phenylalanine and tyrosine. Functional immaturity has been found in both the phenylalanine hydroxylating and the tyrosine oxidating systems, and hyperphenylalaninemia in association with hypertyrosinemia is common in premature infants, especially when fed high-protein diets. The enzymes of the urea cycle are present in the human fetal liver at an early stage of development, but the activities are considerably lower in the immature infant than in the adult or full term infant (4); and the urea synthesizing system of the immature human infant appears to show very little capacity to expand when the protein load is increased (5). Thus, hyperammonemia may easily develop in premature infants.

Recent evidence from a variety of sources has suggested that the amino acid imbalances and hyperammonemia may be toxic for the developing central nervous system (6, 7, 8). In the light of the metabolic immaturities found in the amino acid metabolizing systems it seems potentially equally harmful to administer too much protein as too little.

PROTEIN IN THE FEEDING OF PRETERM INFANTS

There is almost universal agreement that the ideal feeding for the infant born at term is human breast milk. The least that can be said, biologically, is that man has survived thousands of years on this formula, which has evolved to contain only about one percent of protein.

Controversially, however, there still exists doubt regarding the amount of milk protein that should be fed to infants of low birth weight, and since prematurely born infants must have survived only rarely, there is no evolutionary evidence that human milk is the optimal nutrient for preterm infants who normally should be fed through the placenta. Human milk with its low protein content was, however, widely used also for preterm infants until 1947 when Gordon and coworkers (9) suggested that premature infants who were fed with a skimmed cow's milk formula at up to 6.0 g of protein per kg per day gained weight more rapidly than those fed on human milk which supplies about 2.0 g of protein per kg per day. Later it was shown by Omans (10), using formulas based on cow's milk protein, that premature infants who were fed less than 2.5 g of protein per kg per day gained poorly, while those who were fed between 3-8 g per kg per day gained equally well. In 1967 Davidson and coworkers (11) concluded that premature infants needed at least 4 g of protein per kg per day in order to show high rates of weight gain. All formulas used, however, were based on cow's milk protein. In 1969 Snyderman (12) observed that premature infants who were fed 8 g per kg per day of cow's milk protein retained more nitrogen than those who received only 2 g of the same protein per kg per day, although the weight gain was similar in all groups. Some other studies have also shown that infants fed with diets containing more protein grow faster, at least in terms of weight, than those fed an amount of protein more comparable to that found in human milk. However, all studies have used cow's milk based formulas (13, 14). Based on the above mentioned studies many low birth weight infants have been fed with relatively high-protein cow's milk formulas for the last decade. The important question as to how the quantity and the quality of the ingested protein affects growth and metabolism of the immature infants has not been adequately studied.

The proteins of cow's milk and human milk differ from each other quantitatively as well as qualitatively. Cow's milk contains about 3 g% of casein-predominant proteins whereas human milk contains only about 1 g% of whey-predominant proteins. Casein and whey proteins differ considerably from each other in amino acid composition. Casein contains much more of the phenolic amino acids phenylalanine and tyrosine than the whey proteins and, in addition, casein has a very low content of cystine. In contrast, the proteins of human milk with its large whey-protein constituent have the highest cystine to methionine ratio of any source of animal protein. In the light of these differences between cow's milk protein and human milk protein, and in the light of the metabolic immaturities in amino acid metabolism in preterm newborn infants, it would be of importance to know the quantity and quality of protein which can provide enough amino acids to sustain a satisfactory rate of growth in preterm infants without overloading their metabolic capacities.

Barness and coworkers (15) compared two milk formulas which differed in quality of the protein, one containing more whey proteins and the other, more casein. Both formulas were low in total protein, containing only 1.5 g%. These diets were, however, not controlled for ash content and were fed ad libitum. No difference could be observed in the rate of weight gain in 152 premature infants. Body weight gain is not the sole nor the best criterion for the nutritional requirements of premature infants and doubt has been presented whether to grow faster and heavier necessarily means to grow better.

Our own studies (16), in which the effects of changes in the quantity and quality of milk proteins have been examined in relation to growth and metabolic homeostasis in low birth weight infants, have clearly shown that high-protein, casein-predominant diets produce marked metabolic imbalances without increased rates of physical growth. No differences were found in the rate of weight gain from the day of regaining birth weight to discharge from the hospital at 2400 g among the different diets; neither was there any significant difference in the rate of gain in crown-rump length or rate of growth of head circumference in use of breast milk versus the different test diets.

Although no differences were observed in the growth parameters studied, striking differences were observed between the different diets when biochemical parameters were compared. Blood urea nitrogen levels were very strikingly elevated in the infants on the high-protein diets and hyperammonemia, and late metabolic acidosis was seen in the infants receiving high-protein, casein-predominant formulas.

The most striking findings were seen in the plasma concentrations of the essential amino acids. The high plasma concentrations of the essential amino acids were clearly a function of the higher protein intake and the urine excretion of amino acids paralleled closely the plasma concentrations. The plasma concentrations of the phenolic amino acids phenylalanine and tyrosine were especially increased in relation to both quantity and quality of the ingested protein. Ninety-eight percent of the plasma phenylalanine determinations of infants on human breast milk were less than 8 μ moles %. It is only on the high-protein, casein-predominant formula that some infants have extraordinarily high concentrations of phenylalanine. About 5% of the phenylalanine values were between 20 and 100 μ moles %, and in a few cases they were higher than 100 μ moles %, which would correspond to 15 to 20 mg% phenylalanine. These concentrations are similar to those found in phenylketonuria. The high concentrations persisted in many cases for several weeks and were highest in the most immature infants. Even more dramatic differences were seen in the plasma concentrations of tyrosine. In infants fed with breast milk or with whey-predominant low-protein formula none of these metabolic imbalances could be observed.

CONCLUSION

Metabolic immaturities in low birth weight infants affect both synthesis and degradation of amino acids. This has to be considered in the planning of feeding regimens for prematurely born infants. Recent studies indicate that low birth weight infants, when fed a diet relatively high in casein-predominant protein, show late metabolic acidosis, have high blood urea and ammonia concentrations, and in addition have high plasma concentrations of essential amino acids and tyrosine. In some individual cases the plasma levels are very markedly elevated for several weeks and correspond to values found in cases with inborn errors of amino acid metabolism. In clinical practice it would be impossible to know which infant develops severe amino acidemias unless plasma amino acids are determined in all cases, and this is practically impossible. These metabolic imbalances may be harmful for the developing central nervous system and are not found in any infant on human breast milk or whey-predominant low-protein formula. Since no differences are found with regard to rates of physical growth among the different formulas it is suggested that casein-rich high-protein formulas should not be used in the feeding regimen of preterm infants.

REFERENCES

1. Dobbing J: The later development of the brain and its vulnerability. In Davis JA, Dobbing J (eds): Scientific Foundations of Paediatrics. London, Heinemann Medical Books Ltd., 1974.

2. Winick M: Cellular growth in intrauterine malnutrition. Pediatr Clin No Amer 17:69, 1970.

3. Gaull G, Sturman JA, and Räihä, NCR: Development of mammalian sulfur metabolism: absence of cystathionase in human fetal tissues. Pediatr Res 6:538, 1972.

4. Räihä, NCR, Suihkonen J: Development of urea-synthesizing enzymes in human liver. Acta Paediat Scand 57:121, 1968.

5. Räihä NCR, Kekomaki M: Development of amino acid metabolism in the human. In Ghadimi J (ed): Total Parenteral Nutrition: Premises and Promises. New York, Wiley and Sons, Inc., 1975.

6. Menkes JH, Welcher DW et al: Relationship of elevated blood tyrosine to the ultimate intellectual performance of premature infants. Pediatrics 49:218, 1972.

7. Goldman HI et al: Late effects of early dietary protein intake in low birth weight infants. Pediatr Res 7:418, 1973.

8. Mamunes P et al: Intellectual deficits after transient tyrosinemia in term neonates. Pediatr Res 8: 344-370, 1974.

9. Gordon HH, Levine SZ, McNamara H: Feeding of premature infants: a comparison of human and cow's milk. Am J Dis Childr 73:442, 1947.

10. Omans WB et al: Prolonged feeding studies in premature infants. J Pediat 59:951, 1961.

11. Davidson M, Bauer CH, Dann M: Feeding studies in low birth weight infants. J Pediat 70:695, 1967.

12. Snyderman SE _et al_: The protein requirement of the premature infant. I. The effect of protein intake on the retention of nitrogen. J Pediat 74:872, 1969.

13. Babson SG, Bramhill JL: Diet and growth in the premature infant. J Pediat 74:890, 1969.

14. Goldman HJ _et al_: Clinical effects of two levels of protein intake on low birth weight infants. J Pediat 74:881, 1969.

15. Barness LA _et al_: Progress of premature infants fed a formula containing demineralized whey. Pediatrics 32:52, 1963.

16. Räihä NCR _et al_: Milk protein quality and quantity: biochemical and growth effects in low birth weight infants. Pediat Res 9:370, 1975.

INDEX